Ethnicity, Gender and the Border Economy

For whom and why are borders drawn? What are the symbolic projections of these physical realities? And what are the symbolic projections of these physical realities? Constituted by experience and memory, borders shape a "border image" in the minds and social memory of people beyond the lines of the state. In the case of the Turkey-Georgia border, the image of the border has often been constructed as an economic reality that creates "conditional permeabilities" rather than political emphases. This book puts forward the argument that participation in this economic life reshapes the relationship between the ethnic groups who live in the borderland as well as gender relations. By drawing on detailed ethnographic research at the Turkey-Georgia border, life at the border is explored in terms of family relations, work life, and intra- and inter-ethnic group relations. Using an intersectional approach, the book charts the perceptions and representations of how different ethnic and gendered groups experience interactions among themselves, with each other, and with the changing economic context.

This book offers a rich, empirically based account of the intersectional and multidimensional forms of economic activity in border regions. It will be of interest to students, researchers, and policy makers alike working in geography, economics, ethnic studies, gender studies, international relations, and political studies.

Latife Akyüz received her PhD from Middle East Technical University (METU) in 2013. She has held positions as a visiting scholar at Indiana University (2010), Binghamton University (2011–12) and the Center for Ethnic and Migration Studies, Liege University (2014). Her research interests are border regions, ethnicity, and gender studies, with her most recent work focusing on migrant Alevi women living in Europe. She was one of the editors of the *Toplum ve Bilim* (Society and Science) Special Issue on Borders and Border Studies in Turkey. She held the position of assistant professor at Duzce University from March 2014 until her suspension in 2016 due to her signing of the Academics for Peace petition.

Border Regions Series

Series Editor: Doris Wastl-Walter, University of Bern, Switzerland

In recent years, borders have taken on an immense significance. Throughout the world they have shifted, been constructed and dismantled, and become physical barriers between socio-political ideologies. They may separate societies with very different cultures, histories, national identities or economic power, or divide people of the same ethnic or cultural identity.

As manifestations of some of the world's key political, economic, societal and cultural issues, borders and border regions have received much academic attention over the past decade. This valuable series publishes high quality research monographs and edited comparative volumes that deal with all aspects of border regions, both empirically and theoretically. It will appeal to scholars interested in border regions and geopolitical issues across the whole range of social sciences.

For a full list of titles in this series, please visit www.routledge.com/geography/series/ASHSER-1224

Russian Borderlands in Change May 2016
North Caucasian Youth and the Politics of Bordering and Citizenship
Tiina Sotkasiira

Transnational Frontiers of Asia and Latin America since 1800
Edited by Jaime Moreno Tejada and Bradley Tatar

The Politics of Good Neighbourhood
State, Civil Society and the Enhancement of Cultural Capital
in East Central Europe
Béla Filep

European Borderlands
Living with Barriers and Bridges
Edited by Elisabeth Boesen and Gregor Schnuer

Ethnicity, Gender and the Border Economy
Living in the Turkey-Georgia Borderlands
Latife Akyüz

Ethnicity, Gender and the Border Economy

Living in the Turkey-Georgia Borderlands

Latife Akyüz

Routledge
Taylor & Francis Group

LONDON AND NEW YORK

First published 2017 by Routledge

2 Park Square, Milton Park, Abingdon, Oxfordshire OX14 4RN

52 Vanderbilt Avenue, New York, NY 10017

Routledge is an imprint of the Taylor & Francis Group, an informa business

First issued in paperback 2018

Copyright © 2017 Latife Akyüz

The right of Latife Akyüz to be identified as author of this work has been asserted by her in accordance with sections 77 and 78 of the Copyright, Designs and Patents Act 1988.

All rights reserved. No part of this book may be reprinted or reproduced or utilised in any form or by any electronic, mechanical, or other means, now known or hereafter invented, including photocopying and recording, or in any information storage or retrieval system, without permission in writing from the publishers.

Notice:
Product or corporate names may be trademarks or registered trademarks, and are used only for identification and explanation without intent to infringe.

British Library Cataloguing in Publication Data
A catalogue record for this book is available from the British Library

Library of Congress Cataloging in Publication Data
Names: Akyèuz, Latife, author.
Title: Ethnicity, gender and the border economy : living in the Turkey-Georgia borderlands / Latife Akyèuz.
Description: Abingdon, Oxon ; New York, NY : Routledge, [2017]
Identifiers: LCCN 2016038482 | ISBN 9781472481863 (hbk) | ISBN 9781315580524 (ebk)
Subjects: LCSH: Hopa (Turkey)--Ethnic relations. | Hopa (Turkey)--Economic conditions. | Hemshin--Ethnic identity. | Laz--Ethnic identity. | Borderlands--Social aspects--Turkey. | Borderlands--Social aspects--Georgia (Republic)
Classification: LCC DR741.H67 A59 2017 | DDC 305.8009566/22--dc23
LC record available at https://lccn.loc.gov/2016038482

ISBN: 978-1-4724-8186-3 (hbk)
ISBN: 978-0-367-13877-6 (pbk)

Typeset in Times New Roman
by Taylor & Francis Books

To the Academics for Peace of Turkey

Contents

Contents

Introduction

We are at a time when many issues in Turkey's political agenda correspond to problems about borders. We watch on TV the destruction created by a war going on in Syria, a war that has recently infiltrated within the borders of Turkey; we watch lost lives, displaced people, and cities getting bombarded every day. The state of Turkey shells Kurdish towns, kills its own citizens, and imposes forced migration on them with the purpose—or under the pretext—of protecting its borders. The European Union, on the other hand, is in negotiations with Turkey for measures and new practices in order to prevent refugees from entering the EU for the security of the areas within its borders. Ironically I, a scholar of border studies, am imprisoned within the borders of this state of which I am a citizen for speaking about these latest developments and criticizing the state.[1] A "ban on leaving the country by a judicial control decision" is imposed on me. My name is reported to all existing border crossings to prevent an illegal crossing. Hence, I am conducting this study at a time when I starkly feel the physical reality of borders. Luckily, I had finished the fieldwork for the study and was able to cross to the other side of the border earlier.

For whom and why are borders drawn? Who are they defined to protect, and from who and what? For what are they the beginning and end? And what are the symbolic projections of these physical realities? Scholars of border studies have been asking all these questions and looking for their answers in different ways for many years. I looked for the answers to these questions at the Turkey-Georgia border. While some of the questions have some explanations valid for all borders, some questions correspond to a different reality in each border area. The answers change depending on whether there is conflict at a border, on political and economic relations with the state on the other side of the border, and on ethnic and historical relations between the border people on either side of the border. The border that is subject to this study—unlike Turkey's borders with Iran, Iraq, or Syria—is one that has remained closed for many years, was reopened after the dissolution of the Soviet Union, and since then has been on the agenda in relation to the trade carried out through the crossing without being subject to military conflict.

Whether there is conflict at a border or not, the relations with the other side of the border shape a "border imagination" in the minds and social memory of the border people. For Turkey's borders too, differences emerge in the ways in which the borders find a place in the minds and social memory of the border people according to the bordering states, the economic and political relations established with the bordering states, the ethnic components on either side of the border, economic relationships, and the cultural similarities and differences. For example, in Gürkaş' study that focuses on southeastern borders where "the state of emergency has become ordinary" Gürkaş (2014), she describes how the image of the border on that geography is constituted by experience and memory and is turned into a "region"[2] in the minds of people beyond the lines of the state. However, as we will see throughout this book, in the case of the Turkey-Georgia border, the image of the border has often been constructed as an economic reality that creates "conditional permeabilities" rather than political emphases. As Green (2014) put it, it is exactly for this reason that describing borders as physical objects is to focus only on the visible part of the iceberg. Yet physical and symbolic meanings of the border are intertwined. This understanding is also an indication of the shift in approaches to border studies. For Alvarez, for example,

> rather than maintain a focus on the geographically and territorially bounded community and culture, the concepts inherent in the borders genre are alert to the shifting of behavior and identity and the reconfiguration of the social patterns at the dynamic interstices of cultural practices... we need to examine paradox, conflict, contradiction, and contrasts.
>
> (Alvarez 1995, 462)

This transformation in border studies can be defined in the most comprehensive way as a shift from a state-centered perspective to a border-centered perspective.[3]

When looking at the state-centered approach (that is to say, when looking at the border from the center of the state), as the borders of the state's sovereignty, we see lines that are unchanging, without the possibility of intervention, and exact. Scholars in this camp view borders primarily as sources of political conflict, and they conceptualize and label borderlanders as potential deviants conducting subversive cross-border activities including smuggling, prostitution, and illegal immigration. A border-centered approach,[4] on the other hand, looks at the border from within the border region and the border. According to this view, borders are not lines that are drawn sharply, unsurpassable, and insurmountable. Scholars in this camp reject the idea that borders control, prevent, and regulate economic, political, and social interactions among the borderlanders. Instead, "borders are both sources of unity and identity, conflict and conflict resolution, and are important in defining and transforming national culture and identity" (Donnan and Wilson 1999). Borders and boundaries are both barriers and opportunities; they simultaneously unite and divide and include as well as exclude. How the border is approached also determines what issues will be taken up and investigated in the border. While subjects like the regions of sovereignty of the nation-state, political borders, citizenship gain weight in the state-centered approach, a border-centered approach focuses on subjects like culture and identity through the interaction that happens in border regions. "By focusing on migration and movement across the border, rather than on what the border contained and separated, a whole slew of hybrids, fluid identities and mixtures would be revealed" (Green 2009, 3).

This new approach to border regions coincided with a specific period and chain of events in history. Dissolution of the Soviet Union toward the end of 1980s and the end of a bipolar world order, borders being opened, shut down, or redrawn, the expansion policies of the European Union, and the increasing mobility of people, money, commodities, and ideas in this New World Order facilitated a discussion of borders with new concepts. The necessity of determining new geographical borders after World War I caused "a boom in border studies and empirical approaches focused on the new reality" (Anderson 1982). In the "'dual world" situation that appeared after World War II, on the other hand, the borders became symbols of political and ideological power in the exact sense and illustrated "cases of conflict among countries" or "curtains' among competitive ideological systems" (Minghi 1991). The fast and easy border/transborder movement of money, knowledge, and people that is experienced throughout the whole world has put the existence of the nation-state up to debate. In particular, the "borderless world"[5] discourse of Ohmae (1990), which is related to the globalization, has provided the basis for the end of nation-state discussions. However, O'Dowd (2010)'s "world of borders" definition and discussion, which can be counted as against this discourse of Ohmae, can be accepted as a new period in border studies. For O'Dowd (2010, 1032), contemporary border studies claims that "the 'era of the nation-state', the successor to 'the age of empire', has now given way to a 'post-national' world – 'beyond the nation-state'". While this literature usefully points to many of the ways in which contemporary state borders are being reconfigured, it tends to obscure and downgrade their multidimensionality, distinctiveness, and globality. According to Paasi (2011, 15), "in spite of globalization and the apparent opening of borders, states still have a great interest in maintaining their relative power in the governance of space economy, the minds and well-being of citizens, and thereby social order and cohesion." These "new relationships between nation-states and global processes are manifested extensively in the borderlands," according to Konrad and Nicol (2011, 70). For this reason, the necessity of describing and discussing this new process and new form of relationship with new concepts has appeared. A lot of borders have emerged as a "third space" between states, as spaces occupied by a heterogeneous and hybrid culture, with re-imagined citizenship, and people who share new visions of community across borders, as geographical visions of transactional, socially constructed, formative space, and distinctive place in this process (Bhabba 1990; cited in Konrad and Nicol 2011, 71). This period has also caused the borders to come forward in the agenda again and in a stronger form and new issues. "Border studies scholars enter into dialogue with all those who wish to understand new liberties, new movements, new mobilities, new identities, new citizenships and new forms of capital, labor and consumption" (Wilson and Donnan 2012, 1).

Borders, of course, had been discussed long before this period, and while most debates have focused on the nation-state, they acquired new perspectives. Studies on the Mexican-U.S. border revolved around the identity-cultural axis, hence discourses about "border people" and "border identities" became a part of the field. From then on, borderlands were no longer conceived as lines determined by the areas of sovereignty of nation-states but as living places where cultures met and entered conflict, where social and cultural identities were reconstructed, and concepts such as culture, identity, place/space were discussed by the way they appeared on the borders. Hence, the debate has shifted from state-centric approaches, which conceive borders as symbols of state control and authority, to border-centric ones, which accentuate the binary structure of borders, at the same time separating and uniting, delimiting and enabling possibilities at once. Beginning from when border studies started, in this book I discuss the Turkey-Georgia border without abandoning the state-centric approach while demonstrating at the same time that border people are not simple implementers of state politics.

The Turkey-Georgia border, as a result of the progress mentioned above, reopened precisely on 31 August 1988 in a period when the Soviet Union broke up and the states began to declare their independence. At the same time, the reopening of the border, as other post-Soviet borders, is a symbol of economic and political change in the world. The reopening of the border revealed the chaos in the economic life, mobility created by thousands of people crossing the border. Groups that met for the first time in the new situation that the border created transformed the space itself. In short, it was a new borderland where all dichotomies[6] about borders could be observed. Borderlands are spaces separated from each other by barbed wire, security points, and watchtowers as well as single units with people living on both sides. The structure emerging on the edge of at least two states produces its own rules, institutions, and way of life, and borderlanders stay between the effects of the political and economic rules of these states. This structure is both determined by sharp lines and rules and is uncertain and ambiguous, and it can be changed easily. The relations established in this structure impose their own rules, finding ways to cope with uncertainty and variability, pushing limitlessness in bordered spaces, and creating a dynamic structure. All organizational structures in borderlands and social and cultural groups are the components of this dynamic structure. Borderlanders who live "in between" regions experience the uncertainty and danger this situation creates, and they experience life in a different way that others. They are on the edge of two or more states and affected by regulatory, economic, and political changes, but they don't belong to either one. That's why they create their own life experience in a liminal space and regulate and establish relations with structures in this way.

A more complete understanding of practices at the border, and an address of the relational dynamics between the actors of these practices, can be possible through analyses of the "new realities" created by the points of intersection of these relational mechanisms. That is to say, the dynamic structure of border regions discussed above emerges as border people, institutional structures, state institutions, and many national and international commercial powers create various relational mechanisms. For this reason, the border should be addressed as a "sociological phenomenon with spatial implications" (Simmel 1997). In the Turkey-Georgia border region, approaching the border merely as a "gate," as the beginning and/or end point for the areas of sovereignty of the two states, and addressing it only by its global contexts lead us to miss the border people as actors of this border, the institutional structures at the border, the economic, political, cultural, and social networks of relations at the regional and local levels. It ignores the reality that creates the dynamism of border areas. Paasi emphasized that borders are sociocultural entities and therefore open to reconstruction with changing sociopolitical conditions. State borders are not merely fixed categories placed between states but are also social, political, and fluid formations. Borders and their meanings depends on time/history (Newman and Paasi 1998, 187), and for this reason, borders are not just lines drawn on the ground. They are "social processes dependent on re-imagining and re-interpreting, therefore they are above all social phenomena" (Houtum 2005, 39). Scholars of border studies have to evaluate their historical development by taking all these mechanisms into account when investigating the structure of the borders they address.

From this point of view, I have grouped the chapters in this book into four parts. Even though it is not a historical research, the first part is allocated for the history of the Hopa border region and the history of the Sarp border gate itself, along with the background of the dominant ethnic groups of the region. This research is based upon 52 recorded (voice recording and note taking) and many unrecorded interviews that were conducted at different times at central Hopa, the border village of Sarp, and the district of Kemalpasa as well as the observations of the interviewer. Since the field study to reach the aim of the study depends on the opinions and experiences of the regional people who participated in the research, this work is planned as a qualitative work that contains sociological and anthropological approaches with cultural, social, and economic dimensions. Oral history, structured in-depth interview, semi-structured in-depth interview, and focus group interview techniques were used together. The data showed me that the opening of the Turkey-Georgia border in 1988 was the result of the dynamics involved in the border studies. The opening of this border created new forms of relations in Hopa's economic, social, and cultural life.

In the second part, I look at how economics have been shaped and examine the dynamics emerging in response to the new relations in economic life since the 1988 border opening. We can pursue the traces of the new dynamics that the border economy creates by starting with the experiences of ethnic and gender groups. For this reason, the third and fourth parts are devoted to investigation of border experiences of ethnic and gender groups. The Lazis and the Hemshins, the fundamental components of Hopa's economic, political, and socio-cultural life, the relationships of these groups, and the new dynamics that the border economy creates in these relationships, are handled in the third part. The fourth part evaluates the relationship that gender establishes with the border economy and the direct or indirect effects of participating in these relationships. The focus is the household, which offers the most essential source of information in reading about these transformations. Separate interviews were conducted with women and men of the same household to identify the effects of the border life over intra-group relationships and family. The perception of the "immigrant woman" becomes a core component of this part when observing gender-based discourses of changing life in Hopa. During the fieldwork, I faced situations that I could not explain only by gender or ethnic division. It was necessary to evaluate situations where ethnicity and gender intersected with the border economy with an intersectional perspective. The conclusion addresses how economy, ethnicity, and gender intersect in border regions and the consequences of this intersection on daily life in the case of Hopa. The idea of intersectionality can yield the types of analysis that can cover "new forms of inequality" in the global world. Recent analysis on intersection of race, gender, and class "has posited the idea of interconnecting divisions: that each division involves an intersection with the others" (Collins 1990; Anthias and Davis 1992; Crenshaw 1994; Brah 1996; Anthias 1998b; cited in Anthias 2005). In this way, classes are always gendered and racialized, gender is always classed and racialized, and so on. Thanks to this perspective, it is possible to read multidimensional forms of inequality that emerge in the Sarp border region through the points of their intersection.

Notes

1 In early January 2016, 1,128 academics published a statement titled "We will not be a party to this crime" demanding that the state of Turkey stop the massacre it conducted in Kurdish towns and return to the negotiations for peace. Following the publication of this statement, we became the target of criminal and administrative investigation within universities. Many academics were fired or suspended. Four academics remained imprisoned for 40 days. As one of the signatories, I was also suspended from my job at the university. Currently there are three separate ongoing administrative investigations. I am banned from leaving the country for the duration of the legal investigation. Several articles describing the process for Academics for Peace are accessible at https://barisicinakademisyenler.net, https://barisicinakademisyenler.net/node/168, http://internationalsolidarity4academic.tumblr.com, and http://mesana.org/committees/academic-freedom/cases/turkey-academicsforpeace.html.
2 The term "region" is a political term used especially by those who do politics in the region and its people to denote "Turkish Kurdistan," encompassing all Kurdish towns in southeastern Turkey.
3 While Alvarez (1995) defines these approaches as "literalist" or "a-literalist," Kamazima (2004) distinguishes as "state-centered" and "border-centered." For Alvarez (1995, 3) these "literalists" have focused on the actual problems of the

border including migration, policy, settlement, environment, identity, labor, and health. The "a-literalists," on the other hand, focus on social boundaries on the geopolitical border and also on all behavior in general that involves contradiction, conflict, and the shifting of identity.
4 For the studies that handle the border with this approach, see Asiwaju and Adeniyi (1989), Martinez (1998), Flynn (1997), Donnan and Wilson (1999), Papademetriou and Meyers (2001), and Baud and Schendel (1997).
5 For Ohmae (1990, 13), the nation-state is becoming obsolete, because it is no longer the optimal unit for organizing economic activity.
6 For Martinez (1998, 305), the border is "predictable and unpredictable; it divides and unifies; it repels and attracts; it obstructs and facilitates," and due to its bipolar environment, it is normal that border people reflect contrary tendencies such as "conflict and accommodation, poverty and wealth, racial animosity and tolerance, and cultural separation and fusion."

References

Alvarez, R. 1995. "The Mexican-U.S. Border: The Making of an Anthropology of Borderlands." *Annual Review of Anthropology* 24: 447–470.
Anthias, F. 1998b. Evaluating Diaspora: Beyond Ethnicity?, *Sociology* 32(3): 557–580.
Anthias, F. 2005. "Social Stratification and Social Inequality: Models of Intersectionality and Identity." In *Rethinking Class: Culture, Identities and Lifestyles*, edited by F. Devine, *et al*. New York: Palgrave Macmillan.
Anthias, F., & Yuval-Davis, N. 1992. *Racialized boundaries: Race, nation, gender, colour and class and the anti-racist struggle*. London: Routledge.
Asiwaju, A. I., & Adeniyi, P. O. 1989. *Borderlands in Africa: A Multidisciplinary and Comparative Focus on Nigerian and West Africa*. Lagos: University of Lagos Press.
Bhabha, H.K. 1990. *Nation and narration*. New York: Routledge.
Brah, A. 1996. *Cartographies of Diaspora: Contesting Identities*. London: Routledge.
Collins, P.H. 1990. Black feminist thought in the matrix of domination. In *Social theory: The multicultural and classic readings*, edited by C. Lemert. Boulder, CO: Westview Press.
Crenshaw, K. 1994. Mapping the margins: Intersectionality, identity politics and violence against women of color. *Stanford Law Review* 43(6): 1241–1279.
Donnan, H., & Wilson, M. T. 1999. *Borders: Frontiers of Identity, Nation and State*. New York: Berg.
Flynn, D. K. 1997. "We Are the Border: Identity, Exchange, and the State along the Benin- Nigeria Border." *American Ethnologist* 24 (2): 311–330.
Green, S. 2009. "Scientific Report to COST on WG4, Documents, Techniques and Technologies." Based on the findings of a workshop held in May 2009 on "Passports and passing: everyday encounters with borders." Available online at www.eastbord net.org/events/2009/workgroups/WG42009/Reports/WG42009_Report.htm.
Green, S. 2014. "Sınır ARAŞTIRMALARI: Alanla Ilgili Bazı Düşünceler." *Toplum ve Bilim* 131: 32–44.
Gürkaş, T. E. 2014. "İştisna Hali Olarak Sınır: Mardin-Kızıltepe İkiz Kampları." *Toplum ve Bilim* 131: 219–236.
Houtum, H. 2005. "The Geopolitics of Borders and Boundaries." *Geopolitics* 10: 672–679.
Kamazima, R. S. 2004. "Borders, Boundaries, Peoples, and States: A Comparative Analysis of Post-Independence Tanzania-Uganda Border Regions" (unpublished dissertation). Minneapolis, MN: University of Minnesota.

Konrad V., & Nicol, H. N. 2011. "Border Culture, the Boundary between Canada and the United States of America, and the Advancement of Borderlands Theory." *Geopolitics* 16: 70–90.

Martinez, J. O. 1998. *Border People: Life and Society in the U.S.-Mexico Borderlands.* Tucson: University of Arizona Press.

Minghi, J. 1991. "From Conflict to Harmony in Border Landscapes." In *The Geography of Border Landscapes*, edited by D. Rumley & J. Minghi. London: Routledge.

Newman, D., & Paasi, A. 1998. "Fences and Neighbours in the Postmodern World: Boundary Narratives in Political Geography." *Progress in Human Geography* 22: 186–207.

O'Dowd, L. 2010. "From a 'Borderless World' to a 'World of Borders': 'Bringing History Back In.'" *Environment and Planning D: Society and Space* 28: 1031–1050.

Ohmae, K. 1990. *Borderless World: Power and Strategy in the Inter-Linked World Economy.* New York: Harper Business.

Paasi, A. 2011. "A Border Theory: An Unattainable Dream or a Realistic Aim for Border Scholars?" In *The Ashgate Research Companion to Border Studies*, edited by D. Wastl-Walter. London: Ashgate.

Papademetriou, D., & D. Meyers. 2001. *Caught in the Middle: Border Communities in an Era of Globalization.* Washington, DC: Carnegie Endowment for International Peace.

Simmel, G. 1997. "The Sociology of Space." In *Simmel on Culture: Selected Writings*, edited by D. Frisby & M. Featherstone. London: Sage.

Wilson, T., & Donnan, H. 2012. "Borders and Border Studies." In *A Companion to Border Studies*, edited by T. M. Wilson & H. Donnan. Oxford: Wiley-Blackwell.

1 From empire to nation state

History of the region and the Sarp border gate

The history behind the border

Hopa, a town of strategic importance by virtue of being located at the Turkey-Georgia border, is located in the eastern part of the Black Sea region. Hence, it can be reached through land, sea, and airline transportation. In the introductory document issued by the government of Hopa, the place is described as being situated

> in the eastern part of the district of Georgia Republic, in the western part of Arhavi, in the southern part of Borkça and in the northern part of Black Sea. Distance of the district from Sarp Border Gate by which the transition to Georgia Republic is provided is 18 km and from the City Center is 65 km. Hopa is in the intersection position on the international highway which interconnects Trabzon, Rize, Artvin, Ardahan, Kars, Erzurum and Georgia Republic to each other.[1]

Hopa was included in the territory of the Ottoman Empire at the beginning of the sixteenth century with the military expedition that Mehmed the Conqueror launched over Batumi. In 1519, Trabzon became an independent province with Batumi in its borders (Kırzıoğlu 1976, 89). Hopa was part of the Çıldır province, founded after Lala Mustafa Pasha's conquest of 1578, until Ahıska was given to the Russians in 1829. After that date, Hopa was made dependent on the Trabzon province (introductory document issued by Hopa District Governorate).

Evliyâ Çelebi Seyahatnâmesi travelled this region in 1640. His view on Hopa (at the time known as Hoban) gives us a description of the population. He states,

> It is a beautiful place which is abounding in vineyards and gardens along the seashore and bounded to Trabzon territory. Almost all people living in here are Çağatay Laz and a small part of the population is of Greek origin.

The name Hopa is derived from the words *Hub* and *Khub* in Persian and means "beautiful, charming place" (Seyahatnâmesi 2011, 97).

Until the Ottoman Empire, Hopa was part of Lazika. The seashore beyond Rize was called Lazistan in the Pontus Empire and the Ottoman Empire. After it was conquered by Ottomans, Turks moved into the region, especially to places where there were fertile grounds and farms, as Çelebi mentions in his travelogue.

Both Halit Özdemir's *Artvin Tarihi* (*History of Artvin*) (Özdemir 2001) and Nebi Gümüş's unpublished doctoral dissertation, "XVI. Asır Osmanlı-Gürcistan İlişkileri" (Ottoman-Georgian Relations in the Sixteenth Century), indicate that at the beginning of the sixteenth century, Prince Yavuz Sultan Selim came to Melo (a central village in Artvin, currently called Sarıbudak) descended on the Castle of Gönye, and conquered the regions of Arhavi, Viçe (Fındıklı), Atina (Pazar), Hopa, Gönye, Batum, Chala, Beğlevan, Noğedi (Kemalpaşa) and Sarp. These regions surrendered to Ottoman sovereignty after the negotiations that followed a very long war.[2] Recai Özgün's *Lazlar* states that Yavuz Sultan Selim conquered Melo together with settlements like Hopa, Gönye, Batum, Arhavi, Viçe (Fındıklı), Atina (Pazar), Çhala, Beğlevan, Makriyal (Kemalpaşa) and Sarp following the battle of Çaldıran (1514). Considering these historical statements, it is safe to set the early sixteenth century as Hopa and its surroundings' inclusion to the Ottoman Empire by Selim I.

The 1869 Trabzon annual census indicates that the male population of Hopa district was 4,496. It also shows 1,261 households in the Hopa district (including its villages), 27 Islamic schools with 925 students, 27 madrasah teachers, 45 orators, 31 imams, 17 masjids, and 3 madrasahs of science (ilmiye) with 80 students. Following the separation of Ahiska from the Ottoman territories and its surrender to Russians as a result of the Edirne Treaty in September 1829, the Hopa district (with its Arhavi and Gönye regions) that was within the state of Trabzon was placed under Batum county with Batum as the center. In 1865, the province system was established in the Ottoman Empire instead of the state system, and as a result of this reorganization, Atina, Hopa, and Hemşin districts were made regions and placed under Arhavi, which was turned into a district center with an arrangement made in 1867. In 1869, the district governorship of Arhavi was carried out by Kapicibasi Suleyman Bey, and a municipality organization was established on this date. Three years after this shift in the center, Hopa was given the status of a district with an alteration in the governing structure, and Arhavi was turned into a region and made part of the Hopa district. In 1871, a municipality structure was established in Hopa.

Part of the Hopa region remained under Ottoman rule until their surrender to the Russian Empire (following the war between the Ottoman Empire and the Russian Empire in 1877–78). Kars, Ardahan, Doğubeyazıt, and the adjacencies of Batumi were given to Russia as war compensation. Doğubeyazıt was taken back from Russia with the Berlin Treaty of 1878, yet Kars, Ardahan, Batum, and what we call today Artvin, Ardanuç, Borçka, Şavşat, which were districts bounded to Batumi in the meantime and also Kemalpaşa district from Hopa, were left to Russia. The Turkish-Russian Bbrder defined immediately after the Berlin Treaty went through Artvin Mountain, Melo (Sarıbudak), Orcuk

(Oruçlu) Steep, Down Hod (Maden), Erkinis (Demirkent), the southeastern plateau hills, Tavusker, and Oltu (Çoruh 2008).

In this way, the settlements of Limanköy, Kemalpaşa, Osmaniye, Karaosmaniye, Köprücü, Dereiçi, Kayaköy, Çamurlu, Üçkardeş, Kazımiye, and Sarp, which are situated in the east of Esenkıyı village today, remained under the occupation of Russia until the Proclamation of the Republic (Koday 1995, 113).

According to the Seventh Article of the Berlin Treaty, "The inhabitants of the abandoned neighborhood units (three districts) in Russia are autonomous in terms of leaving there by selling their properties in the event that they want to reside in the places apart from these countries" (Özder 1971, 78). For this, three years were allowed, and it was said that the ones who had not left there by selling their estates at the end of this period would remain in the citizenship of Russia.

Both in the time of war and dating from the last treaty, a good number of people immigrated to Anatolia from the neighborhoods of Batumi, Acara, Artvin, Borçka, Ardanuç, and Şavşat. These immigrants, their number not exactly known, settled in Samsun, Çorum, Tokat, Yozgat, Adapazarı, İzmit and especially Bursa City. They were called "93 Immigrants" (Özder 1971, 78).

The people who had been in the center of Artvin, Borçka, Şavşat, Ardanuç, and the Kemalpaşa region of Hopa, which had fallen under the hegemony of Russia, led a poor life under Russian rule. In a similar way, the ones settled in Arhavi, the center of Hopa, and Yusufeli regions that remained in Ottoman territory also led a poor life under the repression and tax-related persecution practiced by local seigneurs. Şakir Şevket, who wrote about the history of Trabzon in 1877, stated that the growing crops and products in Hopa and Arhavi districts were not abundant and precious enough to be sold as in the Pazar (Atina) district. He asserted that the people there engaged in agriculture and commerce and were famous for swordsmanship while he made mention of Hopa and Arhavi districts (Şevket 1877, 47).

Turkic people who lived in this area under Russian domination for 43 years until the Kars Treaty in 1921 had not been conscripted and had no opportunity to receive education, but they were free in terms of their religious activities, clothes and apparel, and agricultural activities. For this period, there are not only demographics pertaining to Artvin City but also Russian statistics. According to the statistics of 1917, the ethnic division of the 985,000 Cenüb-i Garibi Caucasian people living in Batumi, Kars, Ahıska, Ardahan, Artvin, Ardanuç, Oltu (including the Şenkaya district), Kağızman, Iğdır and Azgur was as follows:

Muslim (Turkish): 700,000
Armenian (most of Aras and Arpaçay tribes): 200,000
Greek (all immigrants): 40,000
Russian (all settled in the north part of Aras and Kor River as immigrants and clerk-soldiers): 30,000
Georgian: 15,000 (Çoruh 2008)

Nobody owned the right of disposition over the land. Land was given within definite limits and on behalf of the village; the work of land allocation was carried out by the villagers. Land was apportioned once in three to five years in accordance with the population ratio, and disputes arising in the course of this arrangement were resolved by the government. The villagers did not have the right to change and sell the allocated land. Yet the lands reserved for vineyards and gardens could be used as real property and were not included in the scope of periodic allocations.

Rize had been a *mutasarrıflık*, the local government of the *sancak* (a sub-division of a province in the Ottoman administrative sections). It became a province on 20 April 1924. Hopa remained part of the Rize province until 4 January 1936. After that date, it was part of Artvin. According to the results of the 1927 census, the population was 31,080 including Arhavi, which was a sub-district of Hopa in that period, aside from Fındıklı and Kemalpaşa.

Because it was located on the shore, Hopa's commercial dynamism was more or less comparable to the inner parts. The clearest indication of this dynamism was the population. For instance, according to the census in 1927, the population was 687 in Şavşat and 1,704 in the center township of Artvin, while 4,241 people resided in the central part of Hopa. This difference is explained by the maritime and commercial activities in the city center (Zeki 1927, 135–39).

The economic dynamism that the port of Hopa created continued with the beginning of tea cultivation, and it reached a whole different level with the reopening of the Sarp border gate in 1988. Even though Hopa continued to attract migrants, the population did not change much, though there were transformations in the demographics.[3] According to information provided by the director of the census bureau, it received migrants mostly from Rize and Ardahan. Most of the people from Rize moved to Borçka and from there to Hopa, usually because of blood feuds. They were told "not to come." The migrants from "the east" come in order to work at sharecropping and construction jobs and then settle down. There is no record of the numbers of different regional or ethnic identities within the region. However, the greater part of this population is made up of Lazi and Hemshin. Lazis are the primary group from Hopa that migrated primarily to İstanbul and Zonguldak. Migration of the Hemshin from villages, and because they had more children than the Lazis, equated to the Lazi and Hemshin populations at the Hopa center.

Sociodemographic composition

Until the establishment of the Turkish Republic, the region from the eastern border of Trabzon to the inner sections of Georgia was called Lazistan. Even today, the people who live in this seashore region, which includes Hopa, are overwhelmingly Lazi. However, there is an important Hemshin population in central Hopa and the Kemalpaşa district. The Hemshin who came to the region in the sixth century, settled in the counties of Hemşin and

Çamlıhemşin, and spread throughout the region resided in mountain villages near the summer ranges (Aksu 2009, 31). After Lazistan was brought under Ottoman rule, a lot of Turkic clans settled in the region. Along with Ottoman rule, Islamification in the region started. The wars between the Ottoman and Russian Empires and the oppression of ethnic cultures in Caucasia caused huge migration waves. Because the routes immigration used passed through Lazistan, there were many groups with Caucasian origins who lived in the region, but there isn't adequate information about these groups. Today, the biggest groups in Hopa are the Lazi and the Hemshin.

The Lazi

The word "Laz" has been connected with various regions and people of the eastern bight of the Black Sea coast since the early Christian era (Meeker 1971, 320). Meeker differentiates between the words "the Laz" and the ethnic "Lazi." "The Laz" should not be understood as necessarily referring to a specific ethnic or linguistic group. This situation is further complicated by the fact that the Black Sea people who call themselves "the Lazi" and are referred to as "the Laz" by outsiders represent a specific ethnic group and speak a language of their own. The principal settlements of the ethnic Lazi are found today at the extreme eastern end of Turkey's Black Sea shore in the coastal lowlands between Pazar and the Çoruh River. Their language is closely related to Mingrelian and more distantly related to Georgian and Svan. The ethnic Lazi constitute a very small minority, even among Turkey's eastern Black Sea people. The category Laz, as used by Anatolian Turks, does not precisely designate the ethnic Lazi but frequently refers to all Black Sea people of Turkey, and it typically designates the people living along the eastern shore.

Before discussing who the Lazi are and their history, it is necessary to state that in this book, by following Meeker's distinction between "the Laz" and the ethnic "Lazi," the concept of "Lazi" will be used. This concept refers to an ethnic group that lives in the east of the eastern Black Sea region, from Trabzon's east to the Çoruh River, and speaks the Lazi language.[4]

The first thing that comes to mind when "Laz" is mentioned is someone who lives in the eastern Black Sea region and the "Temel" and "Fadime" stereotype of the Black Sea region jokes. According to this stereotype, everyone from the Black Sea region who is Lazi speaks a heavily accented Turkish, has a quick temper, is extremely naive so as to be the subject of jokes, but at the same time is very cunning and hard working. Such stereotypes had negative effects on the region's people, and they denied their ethnic identity. That might be one reason to acquire knowledge of the ethnic plurality of the region, not to mention the nation state's policies of cultural assimilation. The efforts to learn and promote Lazi culture and language that started with music in the 1990s have continued with the endeavors of associations and organizations that challenge this understanding. Nonetheless, there are no systematic academic studies of the history and homeland of the Lazi.

So who are the Lazi? According to Hann and Hann (1995, 488), the Lazi are of Caucasian origin. They preserved their original language, Lazuri Nena, which is related to Georgian and Mingrelian. Although no official statistics on the number of Lazuri speakers are available, it seems unlikely that there are more than 250,000 speakers.

Lazistan as a political and administrative unit has had varying boundaries throughout history. However, the region where the majority of Lazuri speakers live is limited to a much shorter stretch of the eastern Black Sea coast, i.e., between the border village of Sarp in the east and the village of Melyat in the sub-province of Pazar in the west (Hann, C. and Hann, I. 1995, 488).

Hann (1995, 491) indicated that he sees Bryer's (1966; 1980; 1985), Meeker's (1971), Toumarkine's (1995), and Feurstein's (1983; 1992) studies on Lazuri language and Lazi culture as "objective history" writing. He distinguishes these studies, because there are contradictory understandings and writings on the Lazuri language and the Lazi culture.

One of them is the group led by Kırzioğlu (1973, 1976), who claims that Lazis are among the Turkic people who came from Central Asia. The other is an understanding laid out in *Lazların Tarihi* (*The History of Lazi*). It was written by Muhammed Vanilişi and Ali Tandilava from Sarpi village and claims that Lazis are of Georgian-Megrel origin. According to Hann (1995, 495), close genetic and linguistic ties with Megrel and Georgians, which the book argues, are valuable historical evidence compared to Kırzioğlu's work. However, an uncritical selection of sources that are full of nationalist discourse and prejudice makes Vanilişi and Tandilava's work closer to that of Kırzioğlu. According to Öztürk (2005), the politically motivated propositions like Evliyâ Çelebi's (seventeenth century) mixing of Lazis with Lezgis who are an Eastern Caucasian clan, perpetration of this even by the modern Turkish historians who use him as a source; the effort by some historians in order to prove that "Lazis are Turks" (Gologlu, 1973; cited in Öztürk) is far from the truth.

As Öztürk states in the *Black Sea Encyclopedia* (2005), *Lazi* as an ethnic word was first mentioned in Pliny's work, *Naturalis Historia* (cited in Öztürk, 2005, 753). As Procopius (Öztürk 2005, 756) also proposed, it must be a name that identified more than one Colchis clan. The name of this clan that ruled the Colchis people might have evolved from the word *lazani*, which means "the land of strong/noble; those who live in the mountain." The ancient region of Colchis spreads from west Georgia to northeast Turkey, yet little is known about the history, people, and language of the Colchis. The exact founding of Colchis is not traceable. The kingdom of Colchis is mentioned in ancient chronicles since the middle of the sixth century BC (cited in *Corpus Fontium Historiae Byzantinae*) (Sökmen 2008, 4).

The Scottish writer Neal Ascherson (1995, 199) wrote a book based on a trip that he took to the Black Sea region to ask this question.

> Who are the Lazi is to be at once lost in the chaotic building-site of nationalist definitions. The language, Lazuri, is a survival from a

previous, almost lost deposit of human speech. A pre-Indo-European tongue, it belongs to Kartvelian language family of the Caucauses whose other members are Georgian, Mingrelian and Svenatian.[5] Mingrelian is the closest to Lazuri, and it would appear that both peoples were living as neighbours along the eastern shore of the Black Sea as long ago as 1000 BC. This coastal region around the river Phasis, near the modern Georgia ports of Poti and Batumi, was the land which the Greeks called Colchis.

But according to Ascherson (1995, 200), at the some point, a large part of the Lazis abandoned their homeland. They left Colchis and the Caucasus and moved around the southeastern corner of the Black Sea to their present territory in what is now Turkey. The Mingrelians, in contrast, stayed much where they were. Most of them retained their Christian religion, like the Georgians, while the Lazi and the much larger Abkhazian language group living farther north along the Caucasian coast converted to Islam in the fifteenth century. Why and when this migration took place is not known for certain, but it seems to have happened about a thousand years ago in the Middle Byzantine period, and the Lazi may have been displaced by an Arab invasion of the Caucasus.

Lazis were enjoying free reign in their semi-autonomous kingdom in exchange for protecting the eastern border of the Roman Empire. The Lazi, who were in a position as the only ruler of countless small tribes in Colchis, were squeezed between the Persian and Roman Empires. It was not only the two empires that desired to expand their territories and establish their rule over the region, which threatened the Lazi and the foraying nomads from the north. Because the Persian goal was to uproot the Lazi from the region and settle their own people, the Lazi stayed close to Rome (Öztürk 2005).

During the centuries of Byzantine-Persian antagonism, the Lazi seem to have maintained a certain degree of independence. Arab conquest in the late seventh century seems to have put an end to Lazi political independence. The Lazi region was later incorporated into the Trabzon Empire, and even after the fall of Trebizond to the Turks in 1461, the Lazi managed to maintain a certain degree of autonomy. They remained under the rule of their local valley lords (Hann 1995, 489).

Even though the Lazi experienced independence for a short while following the fall of the Trabzon Empire, they converted to Islam en masse under Ottoman rule in 1580. The course of the process of Islamization and Turkicization is not well documented, but something is known of three groups that retained more of the older Pontic culture than the majority of Black Sea Turks. These are the Lazi in the districts of Pazar, Ardeshen, Findikli, Arhavi, and Hopa, the Armenian-speaking Hemshin in the valleys above the Lazi, and the Greek speakers in the old district of Of. All three of these groups are located east of Trebizond, and the villagers of the last two groups are in the upper reaches of the most inaccessible coastal valleys. The Lazi are thought to have turned increasingly to Islam after 1580, and the Hemshin may have begun to take up Islam in the early fifteenth century (Meeker 1971, 340–41).

According to Öztürk (2005), there are three groups who call themselves Lazi today. These are as follows:

1   People whose mother tongue is Lazuri and live in the region starting from the east of the Melyat River in the Pazar district in the Rize province to the Sarp village at the east, including Pazar (Atina), Ardeşen (Artaşeni), Çamlıhemşin (Vijadibi), Fındıklı (Viçe), Arhavi (Arkhabi), Hopa (Xopa), Borçka district as an autochthonous group, and in Sapanca, Akçakoca, Düzce, Yalova, Karamürsel, İzmit and Gölcük as migrants from the War of '93 (the Russo-Turkish War of 1877).
2   The number of Georgian Lazi, the majority of whom migrated to Ottoman lands on 1877, is not known today because of the Georgian government's handling of the Megrel and Lazi within the Georgian ethnic identity.
3   People whose mother tongue is Turkish and who live in eastern Trabzon and Rize's western seashore. Research done on local Turks showed the influence of the Lazi language (Lazuri Nena) and Caucasian tongue on the Turkish spoken in western Rize, Trabzon, and even farther west (Brendemoen 1990; Öztürk 2005).
4   People whose mother tongue is Romaic, almost all of whom were sent to Greece during the 1923 population exchange. Some settled in Russia and ex-Soviet republics during the Ottoman period (Urum/Romeika).

Throughout their history, alongside their courage and bravery, the quarrelsome character of the Lazi was often acknowledged. For example, Ivane Caiani, a Georgian military officer assigned to Borcka, refers to the Lazi who did tobacco sharecropping in Georgian villages in his book titled *Borcka* as such:

> There is nothing but harm that Lazi bring to this country. They don't refrain from harming the place because they are not the settled people of the region. For example, last year, a Lazi has killed a Murgulian soldier in Acarlı. Two hours subsequent to that event, he crossed to the Ottoman Country and he is still on the run. And this year, they have murdered the headman Bolukvadze. One of the culprits was Lazi. He has, like the first one, crossed over to the Ottoman lands. If precautions against these Lazi could have been taken and an end could be put to their bad deeds, it would have been very beneficial and good for the people.
>
> (Caiani 2001, 88–89)

Prior to the twentieth century, the region was never fully integrated into any of the large empires but remained loosely allied to them, serving as a buffer zone. Its modern history somewhat parallels that of classical times: the Persian-Byzantine threats were replaced by Russian-Turkish conflicts (Hann 1995, 489). Like the time of the Roman Empire, the Lazi were placed in this region in order to protect the borders. The Laz displayed particular loyalty to the Ottomans during World War I. In 1924, the Laz *sancak* was abolished as an administrative unit, and the Laz region became an integral part of the Republic of Turkey.

The Lazi are an inseparable component of this border that we are studying, and the history of the border is at the same time their history. However, before delving into the history of the border, we will talk about the other indispensable component of this border: the Hemshin.

The Hemshin[6]

The Hemshin never had an identity as well known as the Lazi, with whom they share a region. Until the movie *Sonbahar* (*The Autumn*),[7] the existence of an identity such as Hemshin or a language like Homshetsma was virtually unknown. We have very little information and documentation about their history and identity because they lived in mountains, away from the city centers in their own world, and stayed out of the academic interest. For this reason, there are a lot of fundamental and unanswered questions about the history and identity of the Hemshin. Do the Hemshin have Armenian or Turkish roots? If they are Turks, then why do they speak an Armenian dialect? If they are Armenian, then why are they Muslim? What are the similarities/ differences between the Hemshin of Hopa and Rize? When and where did they come from?

Elaborating these issues and debates is beyond the scope of this book. However, a brief historical account of this group will be helpful for establishing the basics for a better understanding of the region. Almost all researchers who study the Hemshin have clearly stated that there are two Hemshin groups living in the most eastern part of the east Black Sea region.[8] These two groups are differentiated by language, culture, and territory: Bas Hemshin and Hopa Hemshin.

So far *The Hemshin: History, Society and Identity in the Highlands of Northeast Turkey*, which was edited by Simonian in 2007, is the first and most comprehensive study of the Hemshin. In the introduction, Simonian asserts the following:

> [T]he counties of Çamlıhemşin and Hemşin in the highlands of the province of Rize are the heartland of the now Turkish-speaking western Hemshini, or Bash Hemshini. This group is isolated by the exclusively Lazi county of Arhavi from the Armenian-speaking eastern Hemshini, or Hopa Hemshini, who are mostly settled in the Hopa and Borçka counties of the Artvin province.
>
> (Simonian 2007)

Economically, the Hemshin were originally farmers with a typical pasture and cattle economy. Culturally and especially linguistically, however, the two groups are clearly distinguished from one another today. The eastern Hemshin or the Hopa Hemshin still speak, in addition to the official Turkish, a characteristically western Armenian dialect that is rather different from standard western Armenian. They themselves refer to this dialect as Homshetsma (or Hommecma) or, in Turkish, Hemşince. The western Hemshin, by contrast, speak only Turkish today, and their dialects are somewhat different from those in the area of Hopa (Bläsing 2007, 279).

Linguist Vaux (2007, 257) traces the separation in the use of the language and talks about three Hemshin groups:

1   The eastern Hemshin/Homshetsik, who live in the province of Artvin (with smaller numbers dispersed elsewhere in Turkey, Central Asia, and Europe), speak a language called Homshetsma and are Sunni Muslim.
2   The western Hemshin, who live in the Turkish province of Rize (as well as in larger Turkish cities and Europe), speak Turkish and are Sunni Muslim.
3   The northern Homshentsik, the descendants of non-Islamicized Hemshin Armenians formerly of the provinces of Samsun, Ordu, Giresun, and Trabzon, who live in Georgia and Russia, speak Homshetsma and are Christian.

The date the Hemshin migrated to the districts of Hopa (Khopa, central district) and Makrial or Makriali (the present-day Kemalpaşa district of Hopa county) to the east of Hemşin remains unknown. According to Torlakian, who estimates that 10 to 15 percent of the total population of the Hemshin moved to Hopa, the migration took place during the second half of the seventeenth century. The same approximate period is given by Minas Gasapian. Russian sources indicate a later date of settlement (around 1780 for N. N. Levashov and the early nineteenth century for E. K. Liuzen). Liuzen was told in 1905 by an elderly Hemshinli woman that her ancestors came to the Makrial district a century before (Simonian 2007, 80).

Simonian notes that according to the Ottoman files, the overwhelming majority of the population of the Hamshen province was Christian until the late 1620s. The Islamicization of the Hemshin began in the seventeenth and eighteenth centuries (Simonian 2007; Benninghaus 2007).

In his unpublished article "The Hopa-Hemshin: Social and Political Life," Cemil Aksu states that the Hemshin have undergone three major historical events of disassociation:

1   Their departure from Armenia proper as a result of the first migration toward Hemshin, thus restricting future relations between them and other Armenian communities
2   The Islamicization that began after the Ottoman conquest of the eastern Black Sea region
3   The disconnection resulting from the religious and cultural assimilation stemming from the Turkification and modernization processes implemented by the newly formed centralized Turkish state

All these are components of the "hybrid" Hemshin identity. There are many Hemshin villages in Hopa and Kemalpaşa today: Başoba/Ghigoba, Yoldere/Zhulpiji, Çavuşlu/Chavoushin, Koyuncular/Zaluna, Eşmekaya/Ardala, Güneşli/Tzaghista, Balıklı/Anchurogh, Kaya Köyü/Ghalvashi, Çamurlu/Chanchaghan, Şana, Üçkardeş, Köprücü, Osmaniye, Karaosmaniye/Ghetselan, Akdere/Chyolyuket, and Kazimiye/Veyi Sarp. Usually the Hemshin villages are made up of extended families of brothers. But those of Başoba, Ardala, and Hendek are comprised of different families, leading us to infer that the process of settling took place in different time periods. It is only in the villages of Üçkardeş and Köprücü that the Hemshin live alongside the Laz. Today, the total population of Hemshin is estimated to be around 150,000. According to Simonian (2007),

> The Bash Hemshin are estimated to number around 29,000 individuals in the Rize province, while the Hopa Hemshin are estimated at around 26,000. To these figures must be added the dozen or so villages in the northwestern provinces of Düzce and Sakarya, settled by the Hemshin during the last decades of the nineteenth century, with a population of around 10,000. Large communities of Hemshin are also to be found in regional centres, such as Trabzon and Erzurum, and in the large cities of western Turkey, Istanbul, Ankara and Izmir. Hemshin living in the latter cities probably now outnumber those who stayed in their home villages. In addition, an estimated 3,000 Hemshin live in the former Soviet Union.

Consequently, a total of approximately 150,000 may be given as a realistic estimate.

The Hopa-Hemshin are the only Muslim Hemshin who speak Homshetsma. The fact that the Hopa-Hemshin preserved their language is linked to their isolation. Their continued self-sustaining village life of agriculture and animal husbandry allows them to pass down the language from generation to generation and maintain certain traditions. This situation began to change with the founding of the Turkish Republic when the central government made the teaching of Turkish mandatory. Even those who became merchants in the towns and those who wished to obtain decent jobs were obliged to know the official state language.

Economic history of the region

As Muvahhit Zeki stated in his 1927 work *Artvin Ili Hakkinda Genel Bilgiler*, Artvin's commerce stagnated because of the lack of roads and vehicles. Since the products of the soil were hardly able to meet the requirements of the city, their sales abroad could not be carried out; only fruit, olives, and small amounts of similar produce were exported to Batumi. Imports from abroad were brought from Istanbul via Hopa Harbor, and kerosene was imported from Batumi.

Drapery and haberdashery, powder, cologne, lavender, photograph materials, etc. brought from Istanbul centrally to Borçka by way of smuggling have been exported to Russia. On the contrary, jewelry and gold have been imported from Russia; yet, both importation and exportation activities are not out of harm's way as well as not being steady. Because of that, there have been very rich people besides the ones who have been involved in a bankruptcy and even lost their lives. The arable lands in Hopa district are very few. The production of vegetable is only to the extent which is able to satisfy the self needs of each family rather than being market-oriented. It has not been given weight to the vegetable gardening since the inhabitants are mostly engaged in boating and commerce. Especially the production of hazelnut among the fruits is at the remarkable level. There are lots of hazelnut and orange trees. The significant part of hazelnut, tobacco, orange and apple produced have been exported together with the corn especially for some years.

(Zeki 1927, 135–39)

After the rigid closure of the Sarp border gate in 1937, the most important source of income for the Hemshin who were unable to cultivate their lands was fishing. Until tea production started in 1951, they cultivated just enough vegetables and corn to meet their needs and a limited amount of hazelnuts.

Fishing was the most important source of income for the region before the border was opened. They were catching cod and sturgeon. They were bringing in good money. During the years 1955–60, Russia exercised a sea drill. They used water bombs during these drills. When they dropped the water bomb, this area shook, and since then nets of the angler caught the skeletons. Fish disappeared then. There were various types of fish around Hopa in the Black Sea region; they were all gone.

(ML_OH-1)[9]

Even though they struggled from time to time, the people who lived in Hopa have been better off economically compared to other provinces and districts in the vicinity. When we look at the 40 to 50 years of its history, it is a place where concentrated economic activities take place in some form. Hopa has one of the most important ports of the eastern Black Sea. It has been the access point for places like Artvin, Ardahan, Kars, and Erzurum in its hinterland. These regions have done commerce through Hopa and its port since the olden times. The Port of Hopa was a vital point during the Iraq-Iran war for the development of trade in the region. The transportation sector and tea cultivation have been sources of income for the region since the 1960s. The port reached its full working capacity during the 1970s. Alongside Black Sea Coppers Enterprise and tea factories, the port has turned the town into a lively place.

There were five enterprises in Hopa: Black Sea Coppers, thermal power station, tea factory, port management, and Tekel (tobacco monopoly) headquarters. Now there is only one enterprise left in here. The tea factory belonging to the state has remained. There is no other thing, and those things that were here have been sold or privatized... Now there is only one enterprise, and that is the tea factory. For example, the petroleum office, go and check it out. There is moss covering its door. Its door is locked up; there were 50 people working there, those who left left, and those who didn't leave retired. There is Black Sea Coppers a little farther down; I swear 1,000 people worked there. Now there are two doorkeepers at the gate. Come a little farther down, there is the thermal power station, and 1,500 people worked there. Now there are ten doorkeepers.

(ML_IW-6)

As this interviewee indicated, Hopa had a very active economy between the coup d'etat events of 1960 and 1980. However, the fearful and distrustful context created by the 1980 military coup also affected the economic life and the people of Hopa. This was followed by the privatizing logic of the neo-liberal policies implemented during the Ozal period. Consequentially, state enterprises were either shut down or rendered functionless by being sold to private corporations. This process has caused a lot of people to migrate from Hopa.

Now he has one land slot, the land slot is tiny. What does the father do? When there are three, four children, he divides the land. Than people cannot sustain it, and when they cannot sustain themselves, they are leaning toward other directions. For example, his father was selling ten tons of tea and sustained himself, and he was working in the tea factory, but the kids have grown up, and they have also gotten married. The land slot has decreased. It decreased to three tons, and a man cannot sustain himself on three tons. What is he going to do? He is going to transfer to another place, out of necessity, or she or he will get an education and work for the state that has also come to an end.

(ML_IW-13)

The cultivation of tea has a primary place in Hopa's economy from the 1960s to the 1980s. Chris Hanns' book *Doğu Karadeniz'de Devlet, Piyasa, Kimlik: İki Buçuk Yaprak Çay* (*Two and a Half Leaf of Tea*) elaborates on this in detail. I am going to briefly handle the conditions today through the anecdotes of our respondents.

The cultivation of tea in Hopa

On 24 January 1980, a program called "January 24 Resolutions" was put into effect. The program was an expression of structural transformation in the economic system of Turkey. The most important arrangement decided upon on 24 January was the reduction of the state's portion in the economy, i.e. privatization. Tea cultivation and the process of its marketing were also affected by this. The tea cultivators lost their long-standing customer: subsidies from the state tea monopoly ÇAYKUR.

> The people who have retired from the tea factory, which was here in the 1970s, are holding 30 percent of Hopa's economy. My late father was retired from there, and my mother takes his pension. At least 30 people in our village have retired from there. Look what a good thing it is. But it is over too.
>
> (ML_IW-13)

> Now I am a worker in the factory and a cultivator at the same time. I collect tea from the branches, and of course there is the transportation business. The pulling of the tea to the factory from the locations of purchase is carried out by transportation. There is a transporter cooperative here. If you have a car, I can also earn money with my care. This all means that tea is our backbone. But it has turned really bad with the privatization. Before, if somebody had no source of income, they gave all of it to ÇAYKUR, and he was really able to sustain through that year with that, but now it is not like that.
>
> (FH_IW-3)

For the people who cultivated the tea and worked in tea factories, there were no factories to work in, and no state organization that they could sell the product of their cultivation with trust were left after the privatization. The private factories did not have a contract with the cultivators. The merchants came and took the tea and went, but within a year, they gave money in exchange. There were some off-the-books corporations. The corporations took tea in exchange for food items.

> We are forced to give it to them, because ÇAYKUR implemented quotas on tea with this privatization. This started in 1987–88. As a result of the concentration of these private factories, in order for the producers to be able to give it to the privates, they have started implementing quotas and contingencies. Similar to sugar beets and cotton, together with the imposition of international corporations, following the quota and contingent process that was implemented on our tea, we are left in the hands of these private companies.
>
> (MH_IW-1)

For tea, there is a three-month cultivating process. It starts with cleaning up the tea gardens and readying them with pruning and fertilizing in March and April. It ends in September, when you can harvest three times. The first harvest comes in the first half of May if the weather is warm enough; the second comes at the end of July, and the final harvest is done during September. The harvested tea is taken to the "Place of Purchase" where, following the control of the experts, it is sold to the state within the limits of a quota. The remaining harvest over the quota is sold to private companies. The cultivator is required to sell his or her tea the day of its harvest or the quality of the tea diminishes, and it is impossible for sell it again.

> Tea is not like nuts after all, and we do not have the opportunity to keep it on a side at our house and sell it later. In hot weather, when your tea is taken away from the branch, if you cannot sell it that day, if it stays for the night, the next day the value of that tea diminishes by half. The tea dies, its quality falls, and the tea burns. There is no possibility for tea that is burnt to go into processing; you just destroy your tea that is burnt. What do the people do when they are unable to give their tea to ÇAYKUR? They just give it to whichever private tea factory's car comes and free themselves of it.
>
> (FH_IW-3)

This necessity benefited mostly the private companies. The producers who had to sell off their harvest were forced to accept the prices that the private companies offered. In selling to private companies where there was no supervision, the producers experienced a lot of injustice. They lost some part or the whole of their money to these companies, or instead of money, they had to accept food items whose expiration date had already passed. The tea that was the only source of income for many people for a long time turned into a provisional income rather than the main source of income because of privatization and the decline in the state subsidy. Today, almost everyone who lives in Hopa owns tea gardens; however, the number of people who sustain themselves on the income from tea is very few. The opening up of the border has somehow affected this process too. The people who did not own tea gardens but obtained some income through sharecropping or collecting tea lost this income, because today, Georgian men who come from the other side of the border and collect tea for smaller amounts of money are preferred.

For the people of Hopa who experienced economic difficulty after the privatization of the Hopa port and Black Sea Coppers, the shutdown of state-owned tea factories, and the implementation of quotas on tea, the opening of the border in 1988 became a source of hope.

The history of the Sarp border gate

The historical analysis of borders is especially important for the modern states during the eighteenth and twentieth centuries. In this period, borders all over the world became crucial elements in a new, increasingly global system of states (Baud and Schendel 1997, 214). In Hopa's case, this border was marked by the formation of two relatively new nation states: Turkey at first, and Georgia in the last quarter of the twentieth century. This border, which was set by the Kars Treaty between Turkey and the Soviet Union in 1921, gained different statues at different times.[10] Two important dates after the 1921 treaty are 1937 and 1988. In 1937, the border was closed down, and in 1988, it was opened again. The fifty-one-year period when the border was absolutely closed ended in 1988 as a result of Gorbachev's reorganization (perestroika) and openness (glasnost) policies and the Özal period of neo-liberal political ideology in Turkey.

Baud and Schendel (1997) formulate a life cycle of the borderlands between their formation and dissolution. According to them, the first stage in the life cycle is the infant borderland, which exists just after the borderline has been drawn. Preexisting social and economic networks are still clearly visible, and people on both sides of the border are connected by close kinship links. The border is still a potentiality rather than a social reality. We can define the Sarp border region as an infant borderland from 1921 to 1937. According to Baud and Schendel, the adolescent borderland is the next stage. At that stage, the border is an undeniable reality, but its genesis is still recent, and many people remember the period before it existed. Although economic and social relations are already beginning to be confined by the existence of the new border, old networks have not yet disintegrated and still form powerful links across the border. In the following stage, the border becomes "a firm social reality": this is the adult borderland. Social networks implicitly accept and follow the contours of the border. Cross-border social and kin relations become scarcer and are generally considered problematic.

The declining borderland is the result of the border losing its political importance. New cross- or supra-border networks emerge, often initially economic in character, and these are no longer seen as a threat to the state. The decline of a borderland can be a fairly peaceful process: the border gradually withers away, losing its importance for neighboring states as well as for the population of the borderland.

Finally, they use the term defunct borderland (or the relict boundary, as it is sometimes called) when a border is abolished and physical barriers between the two sides are removed. Border-induced networks gradually fall apart and are replaced by new ones that take no account of the old division (Baud and Schendel 1997, 225).

The process following the reopening of Turkey-Georgia border can be defined as an adult borderland. It has become "a firm social reality."

From 1921 to 1937: Transition period

The Turkey-Soviet Union border was defined with the Treaty of Kars that concluded on 13 October 1921. With this treaty, the creek flowing across Sarp village was accepted as a borderline, and Sarp village was divided into two. But as stated by the interviewers, the people living in the district of Hopa between 1921 and 1937 could drive across the border by means of a document called Transire (*Pasavan* in Turkish), which required only the soldier's signature. They could continue to cultivate their lands on the opposite side. Therefore, the people living in this region could carry out their agricultural activities by passing to the opposite side even if their land had been divided.

> At the beginning, our Sarp village and the village that is called Sarpi on the other side were in fact one village. Of course, when the border was drawn at once, there was the "pasavan" crossing. Both the citizens who live there and the citizens who live in our place have lands on the other side. They go to work, they go to fish, or go to work at Batum. What do they do? Pasavan was given. Pasavan means border crossing. They have given a paper in the shape of an identity. When you are crossing the border, defence officers sign it. They cross, work there, and come back to their houses in the evening.
>
> (ML_IW-13)

The old man that I interviewed indicated that on the Batumi side, 1,400 dunams (346 acres) that belonged to the three families from Sarp village were left, and they harvested crops from this land from the beginning of 1921 to the end of 1936.

When the border line was drawn, although the mosque was in Turkey's side and the imam stood in the "opposite Sarp," they could pass to this side in order to worship on Fridays till 1937. They could cultivate their lands and harvest their crops. But the tension in the Turkey-Soviet Union relationship during the Stalin period in 1937 caused a change, and the border gate was closed to all transitions in one night. Those separated by the border being drawn could not make contact with their families after this date, though they kept in touch with them up to 1937. The ones who wanted to go across, as one interviewee expressed, were obliged to go 1,000 kilometers by passing Kars instead of taking the road that is 200 m. Until 1937, they were meeting their needs for gas oil, vegetable oil, salt, and sugar from Batumi. Between 1921 and 1937, the obligation of issuing a document for someone living in Sarp village was imposed for any export passing over Georgia. This situation provided important economic resources for the people who lived there.

> Before 1937, we supplied our needs from Batumi. We supplied all our needs from the opposite side. Gas oil was attained from there. Salt was attained from there. Vegetable oil (we called it *sivicka*) was supplied from there. Sugar was also supplied from there, so much so that it weighed 3–4 kilos. Like a stone, they were cutting and distributing it. Those days, their way of life was much better than ours.
>
> (ML_OH-1)

Moderate relations up to 1936 could not be maintained in the same way as a result of the Montreux Straits Treaty in 1936. The rights Turkey had on the matter of straits were rearranged in terms of international law and Turkey's association with England.

From 1937 to 1988: The impermeable border

The Sarp border gate was closed to all trespassing in 1937, and this situation brought important changes and difficulties in terms of both economy and social life. After 1937, all kinds of communication were banned with relatives and acquaintances on the other side of the border.

> As boundaries clearly divide neighboring regions and are designed to be the barrier separating inhabitants of the given territory from 'others', mass representations about them are of contrast ('either-or'). This was especially characteristic of totalitarian regimes. In the Stalin epoch, the outside world was pictured as a continuous 'territory of darkness,' from whence originated the threat of war and of enslavement by imperialist countries.
>
> (Kolossov 2005, 627)

> I know from what my grandmother told that my very own grandmother's consanguine sister was on the other side. (My grandmother) said that we went down to the border, and as if like singing a song in Lazuri, we told about our troubles. For example, today we have this in there, how are you, like don't throw this. They call it a Türkü (a traditional form of folkloric song) as if singing of our own conditions. She has said that my sister told me that every day when I drink coffee, I should put my cup in front of the window, and if you don't see my cup there one day, you should understand that I have died. And when my grandmother was telling this, she was crying. She said that whenever I came to the village, the first thing I did was to look at my sister's window, and when I saw the cup, I said, "Good, my sister is alive, she is still living." And she said one day I came and looked from the window when I woke up in the morning. There was no cup or anything, and I understood that day that my sister was dead.
>
> (FL_IW-29)

Serious security precautions were taken around the border checkpoint, the number of the watchtowers were increased, and a field named "tracking field" was formed on the Georgian side in order to determine whether illicit pass was taking place.

> In early days, there were more towers. They had three to four towers, and we had three to four towers. However their security was stricter. The creek formed the border. They installed wire fence right beyond the creek on their area. The height of the wire fence was 2.5–3 meters. They established a tracking field 4–5 meters wide right behind the fence. The tracking field can be explained as follows. The soil is totally scraped off. I

mean there will not be footsteps or anything else. It is softened. When you step on it, footsteps are just visible. We could even know if a dog passed by... Once a month they scraped it off and smoothed it over. They did not leave any footsteps in order to track each and every mark.

(FL_OH-2)

The security measures and the restrictions on the other side of the border were at least as intensive as the ones that existed here. Many of those who were left on the other side of the border were exiled and sent to regions in the interior that are far away from the border.

During World War II, Stalin put ours—and when I say ours, not only the Hemshin but Lazi, Savshati, whoever was there of Turkish origin—they all were sent to exile. They were exiled to Kyrgyzstan and to Kazakhstan. They travelled 20 days by train as my grandmother told it. Of course, then there were many people who died because of hunger and misery, and because it was a time of war, those who went to serve in the army. My father was three years old then when he was exiled from Batum. My grandmother gave birth to a baby, and the baby died of hunger on the train. The Russian soldiers were constantly checking, and they took the dead bodies without any questioning and threw them out of the railway car. And when I say railway car, it was not the normal car in a train. They were carried in freight carriages that are used to carry wood or coal.

(FH_IW-16)

One of the respondents with whom I conducted an oral history interview was 12 during this exile, and he settled in Hopa after the opening of the border. He told me about his days of the exile as such:

During the exile, they gave us 45 minutes, and you took whatever you could during that time. You took whatever you could, and if you couldn't take it, it would remain there just like that. There were cars, and they gave us a car for each house, and we loaded our belongings. They took us and put us on a freight train. They put 11 houses, 12 houses, however many houses that could fit into the train. There they have exiled us.

L: Were many people exiled?

Oooh, how can I say it? How many trains went from here? Three to five trains took off. However, many Muslims were there. They exiled all of them from Georgia, and there were no Muslims left. The name Acur is also Muslim. They wrote that they were Georgians in their passport, and they didn't exile them, but they exiled the rest of them. Hemshin, Lazi, Rizeians, Kurdish—they exiled them all, not even a single heart remained. The men had at least five hundred, six hundred sheep. All of those sheep remained. They had cows, and the cows were left behind too... We rode in a train for 18–19 days.

L: Where did they exile you to?

Central Asia.

L: Did they send everyone to the same place?

Half of them to Kazakhstan, half to Uzbekistan, half to Kyrgyzstan... There were men that hadn't seen their siblings for 30 years. There were times like that.

(MH_OH-5)

Çiğdem Şahin describes the period for this side as follows in her article "Sarp: Destroyed Hidden Paradise" published in *Biryaşam Yerel Tarih*.

[F]orbidden zone sign starts a few miles away from the village border, all trespassing activity is strictly controlled by the military. Except for the local people living in town, no one with the written permission to enter the village is allowed. No one could even drive a nail, take a picture or act freely without informing the soldiers. Especially it was strictly forbidden to make gestures or send some signs when Russian soldiers can see you clearly trough the watchtowers. If you violate this law, Russian soldiers document this through the pictures taken from the watchtowers and send a note to Turkey. Even for a small gesture or sign, protocol meetings lasting for hours were held. Then you get your first penalty of noting that you would be delivered to Russia next time.

(Sahin 2009, 21)

Similar stories were conveyed during our meetings. According to their statements, it was strictly forbidden to look back or point to the other side. When such situations occurred, the soldiers took pictures of the person making gestures or pointing and sent them to the Turkish soldiers. The Turkish soldiers visited the home of the person to warn him or her personally. During this period, the folk song tradition in the Black Sea region had an important function. People liked to send wedding or funeral news to the other side. They would pretend that they were working in an area close to the border and make up a folk song to inform the relatives and acquaintances living on the other side of the border.

We never stop and stare... We never talk. We only said the name and cried. That was how we talked. They looked out of the window without going outside. We were freer. They cracked the window and looked out behind the curtains. It was forbidden to point. There was a creek in between. They were on one side, and we were on the other. We dug out in the fields on our side and sang our folk songs to them. They also spoke the Laz language... While weeding, they were also listening to us. When someone died, they closed the curtains and looked out the window. We could never see someone outside. No gestures were made. I never saw any gestures until 1990. It was extremely hard for them. They had to keep the body in the house for a week.

(FL_OH-2)

Those who married and moved to the other side lost the opportunity to see and talk to their relatives after 1937. In order to pass the border and go the other side, you had to get a document from the governor, get a visa, cross the border over Kars, and travel approximately 1,000 kilometers. For this reason, crossing the border was nearly impossible.

Neighbor of my daughter got married and crossed to this side in 1916–17 before the border was formed. Once it was free, our elderly people went to Batum in order to purchase salt. Then it was closed. Her mother, father, and family were left on the other side. At least 30–40 relatives. Her brothers were on the other side, and she was on this side. Her children were here as well. She could not go, as by then you had to get a passport. They used to travel through Kars. You had to get a document from the governor and a visa from the consulate. To be precise, you had to travel 1,000 kilometers in order to reach a distance of 200 meters.

(FL_OH-2)

Despite all these prohibitions, an active life continued in Sarp village. Guests of Sarp were always there; locals, foreigners, curious people, students, researchers, and all kinds of tourist groups travelled to the village via tours or their own vehicles. On Saturdays and Sundays, the village was crowded with visitors. People were curious about the wooden bridge over the creek separating the USSR and Turkey, which was white on one side and green on the other, representing the flags of the countries. On one side of the bridge, Turkish watchtowers and soldiers were located, and on the other side, there were Russian watchtowers and soldiers (Şahin 2009, 22).

After reopening in 1988

In accordance with the International Territorial Transportation Agreement between Turkey and the Soviet Union, signed on 31 August 1988, the Sarp border gate was opened, and business picked up in the town, bringing a better economy and tourism. Transportation and accommodation facilities were affected positively.

The day was August 31, around 10:00 during the day. The Russian soldiers held the door, and so did ours. But there were, as a guess, at least 15,000 to 20,000 people. When people just mingled, there were no more soldiers. And they allowed it too. Everyone has seen their relatives, their mates, their friends and siblings. In the evening, everyone returned to their own place. After that, we had to pass with passports again.

(ML_IW-13)

As Rautio and Tykkyläinen (2000, cited in Dursun 2007) say, the opening of the borders has not always meant an increase in cross-border economic activities in general. The adverse socioeconomic development and unstable political geography of new countries has pulled down the expectations of neighbouring border areas. Nevertheless, some examples show that entrepreneurs can benefit from opening the borders. Many enterprises found ways to benefit from these new markets in a changing geography in Turkey and Georgia. However, opening the border gate, which was a source of hope in economic terms in those days, were perceived as identical with "hopelessness," "degeneration," and "inequality" due to the social transformations effected.

Today Sarp village is crowded with people as it was in the past. Even more people are coming and going, but they are not curious tourists trying to explore the village. This time, thieves, corrupted people, smugglers, and drug addicts are coming to our village. Truck drivers, van drivers, suitcase-trading women, and white slave traffic are coming and going, bringing all kinds of filth, dirt, and degeneration. Local people are lost in this crowd. In the good old days, everyone knew each other and had close neighborhood ties. People gossiped at the village square, and news spread to the village from there. People used to listen to the radio at the village square, and the news spread through the grapevine. Only the elderly sat on the square; young people played card games or backgammon at the coffee house. Women joined the conversation while passing by the square if it was convenient or just kept going, depending on the situation. Today, with the grotesque shops, restaurants, and other places, our people have lost their old values. They have an "itchy palm" to become rich in the easiest way. Most of the problems emerged during the expropriation of the lands to build coastal road while brothers, sisters, and relatives were sharing the money (Şahin 2009, 23).

Pelkmans (1999, 49), who worked on the other side of the border, mentions the same feelings for the people living on the other side of the border. After fifty years of rigid separation, in 1988, the border between Georgia and Turkey was opened for traffic. At the time, the opening was welcomed on both sides of the border. Many Georgians had relatives across the border with whom contact had been largely impossible since the late 1930s. Many of them took the chance to pay a visit to that other, mysterious world. The opening of the border offered people in Georgia access to "western" consumption and goods, and the opportunity to sell their belongings for hard currency was very valuable in their country at the time. Trade between the countries quickly increased and is still very important for the region as a whole. Although these positive effects seem obvious, the inhabitants of Ajaria tend to describe the new contacts in negative terms.

My informants said the opening of the border caused the spread of diseases and chaos on the markets, and they saw it as a threat to local values. There have been advantages and disadvantages that the reopening of the border brought for people on both sides. While the men emphasize the advantages, the women talk more about the disadvantages. This perspective summarizes the understanding of a lot of our male respondents.

> Disadvantage means taking yourself into a mistake knowingly. The gate is there, and the state is saying "I have opened the gate." Trade is free in Turkey, economy is free, you can do your trade, your whatever, however you like. This is what the state says. Once the state says this, there is nothing that hinders your actions. However, if by saying I am going to be involved in trade, you go to Batumi, and then in different hotels and clubs, you spend your money with this and that. Coming here and to your family, if you say, "I went to Batumi and, you know, Georgia has bedazzled me, done this and done that to me," then this becomes a disadvantage. This is the disadvantage, there is nothing else. It is the humans who create the disadvantage. There is only one disadvantage: nature gets polluted, and the reason for that is that the exhaust gases of the trucks that pass by here pollute our nature. For example, the fruits do not grow as before. There was an abundance of fruits and vegetables before. Now that does not happen. Well, that is a price to be paid. When there is industry, when there is commerce, they come to that city. That means that however much the technology develops, human health is affected, and nature is affected. This is the simple meaning of it.
>
> (ML_IW-13)

As much as an emphasis is made on the economic advantages and the people having benefitted from the economy that the border creates, they are also forced to deal with the chaotic processes of being in a transition zone.

After the border opened in 1988, how people experienced or perceived the chaotic process was studied. From the content of the interviews, it's evident that the experiences of the interlocutors vary according to age, ethnicity, gender, and the distance between their residence and the border. Although age and distance are not considered among the variables, most interviewees had higher negative perceptions. The interviews indicate a wide range of variables that create negative perceptions such as low literacy rates among the youth, environmental pollution caused by the border, noise, and the insecurity of living with strangers. According to one of my respondents, while almost all of the youth who lived in Sarp village when the border was closed were university graduates, this rate dropped considerably following the opening of the border.

> Actually, the literacy rate is very high in our village. However, the youth of 1990 generally do not go to school. You know why? They do not need

to learn how to read and write. They all think that they will make the best money because of the border. Even their fathers do not encourage them to go to school. Money is flowing, so why go to school? They make $1,000 per day, so why to go to school at all?

(ML_OH-4)

The academician and journalist Çiğdem Şahin, who was born in the border village of Sarp, describes the feelings of the village inhabitants.

The worst thing is the village people lost their feeling of living in peace and security in their own town. Villagers have the fear of alienation and become the minority in their own land. In the old days you would meet someone you know when you leave your house, now the village is crowded by the foreigners; villagers used to leave their keys on their doors now theft occurs on daytime now.

(Şahin 2009, 23)

However, it is useful to apply the proceeding separation. For the people who live exactly at the border and those who live a little farther away at the hinterland of the border, perception diverges slightly. The stress that results from the security measures, the commotion experienced at the border gate, and the crowds of people waiting to cross the border have caused the villagers of Sarp to regard the border more negatively.

You would have seen the place where customs is located. You would love the coast. Everything was just covered with linden trees as old as a couple hundred years. They were all cut. Construction started in 1985, and the customs area became fully operational in 1989. It was by the end of 1988, early 1989. How did this happen? They had three or five personnel, and these people were not paid properly. We would bring food from our town and feed customs personnel our own food.

(FL_OH-2)

In the center of Hopa, located 18 kilometers from the border, there is quite a warm perception about the border, especially among the men, because of its "economic returns." In the early days of reopening, living very close to the border had advantages such as getting the smuggled goods first hand, earlier than others, and for a better price. However, in time, as exchange expands to larger areas, this basic relationship cannot be enough to analyse the relations and dynamics created by the border. Therefore, this book is designed to look at the economic relations and sociocultural dynamics that appeared after the reopening of the border in 1988, the effects on diverse ethnic groups and gender relations, and their understanding and experience of the border.

Notes

1 Introductory document issued by Hopa District Governorate.
2 The war is said to have lasted three months without a ceasefire.
3 According to 2010 data of the Turkish Statistical Institute (TUIK), the total population was 32,016 with 17,433 people in the district center and 14,583 in towns and villages.
4 Andrews and Benninghaus (1989, 497) make the same distinction as Meeker. Ildiko-Beller-Hann (1995, 505) also followed the distinction of Meeker and Benninghaus and preferred to use the concept of "Lazi."
5 After lots of discussion and debate about who the Kartvelian are, Humboldt and Krettschmer, European historians of the nineteenth century, came to the conclusion that the Kartvelians were closely related linguistically and culturally to aboriginal peoples of ancient Europe, among them Etruscans and Basques (Benninghaus 1989a, 475–497, cited from Sokmen 2008, 3). Between 2100 and 750 BC, the ethnic unity of the Kartvelians broke into several branches, among them Svan, Zan, and East Kartvelian. That led to the formation of modern Kartvelian languages: Georgian (originating from East Kartvelian vernacular), Svan, Mingrelian, and Lazuri (Kutscher, n.d., cited in Sökmen 2008, 3).
6 They are also called Hemshinli, Hamshentsi, or Hamshenahay.
7 *The Autumn* was filmed by the director Alper Ozcan in 2008 in Hopa, Camlihemsin, and Kemalpasa, and Hemshini, Georgian, and Turkish languages were used.
8 See Hann (1995), Blasing (2007), Simonian (2007), Iskhanyan (2012), and Ersoy (2007).
9 Quotations from interviews are as follows:

> FH_IW-interview number: The interview done with a Hemshin woman.
> FL_IW-interview number: The interview done with a Lazi woman.
> MH_IW-interview number: The interview done with a Hemshin man.
> ML_IW-interview number: The interview done with a Lazi man.
> FH_FG-interview number: Focus group interview done with Hemshin women.
> FL_FG-interview number: Focus group interview done with Lazi women.
> FH_OH-interview number: The oral history interview done with a Hemshin woman.
> FL_OH-interview number: The oral history interview done with a Lazi woman.
> MH_OH-interview number: The oral history interview done with a Hemshin man.
> ML_OH-interview number: The oral history interview done with a Lazi man.

10 Here the concept of status refers to opening or closing the border to traffic either partially or wholly.

References

Aksu, C. 2009. "Sol 'Yerel' ve Sosyoloji Üzerine Bir Deneme: Hopa'da ne Oldu, ne Olmadı?" *Birikim* 240: 24–34.

Andrews, P., & Benninghaus, R. 1989. *Ethnic Groups in the Republic of Turkey.* Wiesbaden: Reichert.

Ascherson, N. 1995. *Black Sea.* New York: Hill & Wang.

Baud, M., & Schendel, W. 1997. "Toward a Comparative History of Borderlands." *Journal of World History* 8 (2): 211–242.

Benninghaus, R. 2007. "Turks and Hemshinli: Manipulating Ethnic Origins and Identity." In *The Hemshin: History, Society and Identity in the Highlands of Northeast Turkey*, edited by H. H. Simonian. Abingdon: Routledge.

Blasing, U. 2007. "Armenian in the Vocabulary and Culture of the Turkish Hemshinli." In *The Hemshin: History, Society and Identity in the Highlands of Northeast Turkey*, edited by H. H. Simonian. Abingdon:Routledge.

Bryer, A. 1966. "Some Notes on the Laz and Tzan (I)," *Bedi Kartlisa: Revue de kartvélologie* (Paris, 1966). Reprinted in *Peoples and Settlement in Anatolia and the Caucasus, 800–1900*. London: Variorum Reprints.

Bryer, A. 1983. *The Crypto-Christians of the Pontos and Consul William Gifford Palgrave of Trebizond*. Deltio Kentrou Mikrasiatikon Spoudon. Athens, p. 22.

Bryer, A., & Winfield, D. 1985. *The Byzantine Monuments and Topography of the Pontos*. Washington, DC: Dumbarton Oaks Research Library and Collection.

Caiani, I. 2002. *Borçka Mektupları*, translated by Fahrettin Çiloğlu. Sinatle Yayınları. Istanbul:Sinatle Yayınları.

Çoruh, P. 2008. Muhacirlik Yılları Yazı Dizisi 5–6. 22 Şubat-11 Mart.

Dursun, D. 2007. "Cross-Border Co-operation as a Tool to Enhance Regional Development: The Case of Hopa-Batumi Region" (unpublished thesis). Middle East Technical University.

Ersoy, E. G. 2007. "Social and Economic Structures of the Hemshin People in Çamlıhemsin." In H. H. Simonian (Ed.), *The Hemshin: History, Society and Identity in the Highlands of Northeast Turkey*. Abingdon:Routledge.

Hann, C. 2003. *Doğu Karadeniz'de Devlet, Piyasa, Kimlik: İki Buçuk Yaprak Çay*. Istanbul: Iletisim Yay.

Hann, C., & Hann, I. 1995. "Myth and History on the Eastern Black Sea Coast." *Central Asian Survey* 14 (4): 487–508.

Hopa Kaymakamlığı Tanıtım Metni. 2008. Available online at www.artvin.bel.tr/hopa-ilcesi.

Iskhanyan, V. 2012. "The Armenian-Speaking Muslims of Hamshen." Available online at http://hetq.am/eng/articles/11632/the-armenian-speaking-muslims-of-hamshen-who-are-they?.html.

Kırzioğlu, M. F. 1973. *Artvin Yıllığı Tarih Bölümü*. Ankara: İçişleri Bakanlığı Artvin Valiliği.

Kırzioğlu, M. F. 1976. *Osmanlıların Kafkas Elleri'ni Fethi*. Ankara: Türk Tarih Kurumu.

Koday, Z. 1995. "Hopa İlçesinin Coğrafyası" (unpublished dissertation). Erzurum Atatürk Üniversitesi.

Kolossov, V. 2005. "Border Studies: Changing Perspectives and Theoretical Approaches." *Geopolitics* 10: 606–632.

Kutscher, S. 2008. "The language of the Laz in Turkey: Contact-induced change or gradual language loss?" *Turkic Languages* 12, 82–102.

Meeker, E. M. 1971. "The Black Sea Turks: Some Aspects of Their Ethnic and Cultural Background." *International Journal of Middle East Studies* 2 (4): 318–345.

Nebi, G. 2000. "XVI. Asır Osmanlı-Gürcistan İlişkileri." (unpublished doctoral dissertation). Istanbul: Marmara University.

Özdemir, H. 2001. *Artvin Tarihi*. Istanbul: Ege Matbaacılık.

Özder, A. 1971. *Artvin Ve Çevresi: 1828–1921 Savaşları*. Ankara: Ay Matbaası.

Özder, A., & Aydın, A. 1969. *Yazı Ve Resimlerle Çevre Incelemesi: Artvin İli 1*. Ankara: Güneş Matbaacılık.

Özgün, R. 2000. *Lazlar*. İstanbul: Chiviyayınları.

Öztürk, Ö. 2005. *Karadeniz Ansiklopedik Sözlük (2 Cilt)*. Istanbul: Heyamola Yayınları.

Pelkmans, M. 1999. "The Wounded Body: Reflections on the Demise of the 'Iron Curtain' between Georgia and Turkey." *The Anthropology of East-Europe Review* 17 (1): 49–58.

Pliny the Elder. 1855. *The Natural History*, edited by John Bostock, M.D., F.R.S. H.T. Riley, Esq., B.A. London: Taylor and Francis.

Procopius. 1914–1940. *History of Wars, The Anecdota or Secret History, On Buildings*. Translated by H. B. Dewing. 7 Volumes. Loeb Classical Library. Cambridge, MA: Harvard University Press.

Şahin, Ç. 2009. "Yok Edilen Saklı Cennet: Sarp. Biryaşam Yerel Tarih, Folklör" *Biyografi ve Coğrafya Dergisi* 6: 17–24.

Seyahatnâmesi, E. C. *1–10 Kitap*. 2011. Yayına Hazırlayanlar: Y. Dağlı; S. A. Kahraman; R. Dankoff. İstanbul: Yapı Kredi Yayınları.

Şevket, Ş. 1877. *Trabzon Tarihi*. Trabzon Belediyesi Kültür Yayınları.

Simonian, H. H. 2007. "Preface." In *The Hemshin: History, Society and Identity in the Highlands of Northeast Turkey*, edited by H. H. Simonian. Abingdon:Routledge.

Sökmen, Y. O., & Chesin, M. 2008. "Essay on the Cultural Preservation of the Endangered Kartvelian Language: Lazuri." *Undergraduate Research Journal for the Human Sciences* 7. Available online at www.kon.org/urc/v7/yuksel-sokmen-2.html.

Tolakyan, B. 1981. Drvagner Hamshenahayeri Patmut'yunits. [Episodes from the History of Hamshen Armenians]. *Banber Erevani Hamalsarani [Bulletin of Erevan University]* 2(14).

Toumarkine, A. 1995. *Les Lazes én Turquie*. Istanbul: ISIS.

Vanilişi, M., & Tandilava, A. 1992. *Lazların Tarihi*. Istanbul: Ant Yayınları.

Vaux, B. 2007. "Homshetsma: The Language of the Armenians of Hamshen." In *The Hemshin: History, Society and Identity in the Highlands of Northeast Turkey*, edited by H. H. Simonian. Abingdon:Routledge.

Zeki, M. 1927. *Artvin Ili Hakkında Genel Bilgiler*. Istanbul: Şavşat Kültür Turizm ve Dayanışma Derneği Yayınları.

2 Border economy in Hopa

New opportunities, new contestations

Do you know what the border is? For example, what does it tell you? It tells you that diesel is very economical and that you are free to pass the border. It becomes profitable to buy diesel across the border and sell it over here. The man sells everything he owns and goes into debt to buy a truck with a tank installed; after a month or so, it stops running suddenly. Now the man is left indebted. After three to five months or a year, the border presents another opportunity. Buying goods from one side and selling them on the other becomes affordable and profitable again. People invest their money in that job and go into debt again. Sooner or later, other jobs end too, and they are left with debts. "There is no certainty around the border; you should never start a business trusting it" (FL_IW-31).

When we start to talk about borders, we start to talk about states, supra-state powers, local micro manifestations of these powers, and most of all, the economy of the border. "Open border" regulations of the economic globalization era that enable free circulation of commodities have rapidly created a security and legitimacy issue. They have led to the "discursive construction of certain categories of people and objects" (Bennafla 2014) and definitions of who and which commodities are acceptable and who and which commodities are objectionable. According to these definitions, the security lines are made bolder. The continuity and future of large economic systems were indexed to policies such as the removal of the borders within the European Union and the creation of insurmountable barriers along its perimeter, implemented at the borders. For this reason, the borders as the perimeters of political, economic, social, and cultural institutions play "key roles in the local, regional, national and international order" (Wilson 2014).

The view of the borders as the perimeter of a whole, as "the new centers and laboratories of social and political change" (Bennafla 2014), and as the focus on their local and regional manifestations are the results of the shift from a state-centered approach to a border-centered approach in border studies. Due to this change of perspectives, "inter-border flows, links and identities" and therefore local components in border areas have also become important. Efforts to identify "the changing terrain of various actors at and across political boundaries" (Wilson 2014) make border studies important for understanding the workings of national, transnational, local, and regional dynamics. The problematic effect of all kinds of legal and illegal border-crossing activities and moving between borders on the identities of people, which is identified by Green (2014, 35) as one of the thematic issues in border studies, will not become clear without focusing on daily economic activities at the borders, their organizational and operational mechanisms, and the actors of these operations.

According to Green (2014), border studies that place people, that is, "the relation between people and space; the politics and control of territories; and the movement of peoples around the world" at the center have not, oddly enough, placed commerce itself at the center of research. However, the study I conducted on the Turkey-Georgia border, and the cases about which you will read below, have shown that identity formation processes at border areas, transformations of group relations of these identities, and relationships that people establish with places are defined by states of inclusion/involvement in economic activities that emerge in these areas. Connected to large economic systems at the macro level, these activities present a picture in which they enable "tax exemption," operate by legal and illegal crossings, and define daily life dynamics of border people on the local level. Focusing on the organization of these crossings, which products are traded, and who the national and local actors involved in the process are enables the understanding of the larger picture that includes both the local and the states.

Like all other border regions, the Sarp border has its own social and economic dynamics. It was not only drawn between two different countries, but it was also a border between two different economic systems. The Sarp border gate has become an opening toward the West for Georgia, which has abandoned a socialist economy and turned her face toward capitalism just like many other post-Soviet countries. With the dissolution of the Soviet Union, many countries that declared independence, including Georgia, entered a period of economic hardship. The opening of the Sarp border gate rendered visible many aspects of this transformation and its tragic results: the "Russian bazaars" that were established just after the opening of the gate, the "suitcase trade," and the entertainment sector, which relies on gambling and prostitution. These examples can be seen as symptoms of the emergence of inequalities and chaos that were created by the transition to capitalism. The trade that is done in the border regions has both legal and illegal dimensions such as smuggling and "billing forgery," which has made its mark in Turkey for a period of time. What has been bought and sold, traded, or smuggled had varied according to the conditions in both sides of the border. In this part of the study, I will focus on these changes in economic life since the day the border was opened, and I discuss the border and the opportunities and disadvantages created by the existence of the border on the basis of the experiences of the local actors of the border economy. As I have noted in the introduction of this book, I will share here quotes from my interviewees when possible, as this study is based on the perceptions and experiences of the border people.

When we compare the times when the border was opened and the present, we see that pictures quite different from each other have appeared within this 28-year time span. Each distinct picture enables us to read the role that state policies, interstate relations and agreements, and the local and regional networks of relations have played in the establishment of economic life and the lives of the border people. Heyman (2014, 17) talks about two main issues on borders. One is the issue of culture production,[1] and the other is the issue of political economy. Heyman focuses on the unequal economic exchange between areas connected by political borders and the inequalities caused by these borders in social redistribution. When we have a closer look at the different pictures mentioned above, it will be possible to view both the inequalities that emerge with the borders that Heyman notes and the intersection points of the issues of economy and culture production.

Selling lives for livelihood: Russian bazaars and suitcase/shuttle trade

When the Sarp border gate reopened in August 1988, thousands of people started to enter Turkey from the countries previously been affiliated with the Soviet Union and started to declare their independence, starting with Turkey's bordering neighbor Georgia and followed by Armenia, Azerbaijan, Kyrgyzstan, Kazakhstan, and Russia. Some of these people crossed the border to find the relatives that they had been separated from and some to travel to other countries, yet the majority crossed the border to sell everything they had ever possessed.

By the time the border was opened, there were two types of exchange. On the one hand, there were those trying to sell whatever they had no matter what the prices were in order to buy as much merchandise as they could. On the other hand, there was a group of people whose activity involved regular visits and the search for the most profitable exchange in the market. At this early stage, street bazaars were the venue for this commercial activity. These bazaars eventually turned into "Russian bazaars" that extended from Hopa to Trabzon. This trade "quickly developed as a form of small-scale and informal entrepreneurship in order to fulfill consumer demands" (Yükseker 2007, 64). The spaces in Hopa immediately reorganized to answer the needs of this new phenomenon as one of my interviewees stated.

> During the time the border was opened, my father was a shopkeeper here. We had a wooden three-story building. My uncle was our business partner, and we were running our house as a hotel. Upper floors were used as a lodge, and the ground floor was a tea house.
>
> (MH_IW-1)

The shops were filled with merchandise that would answer the needs of those who came from the other side of the border. People turned their houses into hotels to provide accommodation for the agents of this commercial activity.

The progression and consequences of the process were different for the countries on each side of the border. Georgia tried to ameliorate the pains of transition to a new economic system by benefiting from the border gate, and Turkey sought to obtain profit by trading with not only Georgia but with other post-Soviet countries such as Kazakhstan, Armenia, and Azerbaijan by providing direly needed consumption goods and services in a period of crisis. Shuttle trade and prostitution were related to each other in Turkey, especially in the port cities.

According to Polese (2012, 22–23),

> a border becomes a source of income in two ways. First, it opens a way to earn income with its role in demand supply chain and second, demand on banned or restricted good increases and this situation creates new opportunities for people on the unrestricted side.

At the Turkey-Georgia border, tax regulations, price discrimination, and banned products created constraints and opportunities for the inhabitants of the region, and they benefited from these potentials of the border. Sometimes smuggling and commercial advantages emerged by tax regulations and agreements between the two countries. Sometimes transportation and brokerage became a source of income, but it was always the service industry that benefited more or less from the border. But the course of proceeding was not the same for both sides of the border. The other side of the border was Georgia, which seceded from socialist regime, had just recently declared independence, and whose economy was in ruins. Georgians whose basic needs were provided by the state (housing, employment, and health just to mention a few) until then lived without knowing what hunger or deprivation meant. After the opening up of the border, they came to Turkey to sell first what they had in their homes and then whatever they could find, which started the Russian bazaars. In this initial period, most of the trade was made by those who came from the Georgian side. During this period, there were two different groups involved in the trade. The first group sold the goods they had at the maximum possible price, bought as much as they could afford, returned to their countries, and stayed out of trade when they had nothing else to sell. The other group crossed the border regularly and traded with a maximum profit in a systematic way. Chelnoki (shuttles), the Russian name given to the traders, brought consumer goods such as apparel and household items in moderate quantities from other countries and sold them in marketplaces in their towns. These activities, also called "suitcase trade" (*bavul ticareti* in Turkish), are usually unrecorded, and they escape legal regulation and taxation in both Turkey and Russia (Yükseker 2007, 61).

The first Russian bazaars were established in Hopa and Trabzon. Later on, small markets were established almost everywhere along the Black Sea coastline. The diversity of goods sold in the Russian bazaars was amazing. Any kind of product was sold from kitchen utilities to furniture, antiques, repair tools, and military stuff to icons of Stalin. After selling their goods in these bazaars, they bought cheaper consumer goods benefiting from tax exemption, then returned to their countries, and earned money by selling these goods there.

> Suitcase trade was very common in that period, it was like, as soon as the gate was opened the people who came set up a market from one end of the coastline till the pier with whatever they had in hand like cameras, dishes, bowls and pots. Women, girls and boys, the whole family with all the children would come and open up tables there, and these lodged at the hotels in the evening. And then, this has led us to seriously, I mean we have seen what dollar meant what Georgian currency meant there and suddenly people have started to talk dollar in that process.
>
> (MH_IW-1)

Ascherson (1995, 196) defines "the country of Lazi" and the Russian bazaars as such:

> [I]n the opposite direction come caravans of old red Ikarusz buses. They list to port or starboard and gush black smoke, like paddle steamers... Their passengers – Russians, Ukranians and people from every nationality in the Caucasus- bring with them anything they can pack and carry – tea-sets and busts of Stalin, toy tanks and lavatory seats, cutlery and clocks, garden furniture and surgical instruments- to sell in the new 'Russian market'...many of the sellers have travelled for days and nights from as far as Kiev or St. Petersburg, paying bribes and protection money at one frontier after another, keeping a special wad of banknotes to sweeten the Caucasian mafiosi who allocate the stalls at Trabzon.

As Roitman (2001) notes, border-specific economic activities usually require a complex organization involving brokerage, patronage, security services, and creative communication and transfer methods. They bring out authority groups competing with the regulatory authority of the state. Every person and every institution from the smallest to the largest involved in these activities in the border area finds themselves in these complex networks of relations. Even someone who takes her or his belongings at home and crosses the border to sell them has to enter into relations with each link of the chain from the person facilitating the transfer to the customs officer at the gate and the small Mafia-type organizations.

These relations show continuity from the departure point of the product to the place where it is sold, and those involved in this trade in this way on either side of the border act as a whole. In this process, not only the areas in which the trade was made but all spaces in Hopa become a trade center. The commodities subject to this trade were products put to sale by those coming from the other side and products bought because they were demanded from the other side, especially in this first period. In this market, in which many different products were sold from guns to small decorations, those who sold these goods were trying to satisfy their fundamental needs. Others bought them either because they were antiques or as decorations for their houses. In a sense, the histories of those who come from the other side of the border became fodder for trade and turned into a commodity that was bought and sold. The Chamber of Shopkeepers and Artisans (pers. comm.) described the process with these words:

> Under normal conditions Georgians were doing suitcase trade here in this coastline, it was the Georgians who have started the first suitcase trade. The daily potential of Hopa did not fall down to 5,000. Even guns were sold, what are those missile things called and police was right besides us and even the police did not know the Kalashnikovs than. The man has normally laid them down to a stall and nobody knows what it is, and nobody has a malevolent intent, he brought them there and is selling, such things as binoculars were very popular.

As the demands and volume of the trade changed, the commodities traded also changed. In time, it was possible to observe how some people who successfully established these networks of relations in border trade turned living on the border into an advantage and created wealth accumulation by integrating their local links with national trade networks. Those who could not achieve the accumulation to sustain these organized relations, on the other hand, were either excluded from this trade or became commodities in this trade as seasonal workers or sex workers.

The economic activities of borders are defined by MacGoffcy (1988; cited in Donnan and Wilson 1999, 88) as "a highly organized system of income-generating activities that deprive the state of taxation and foreign exchange... Some of these activities are illegal, others are legitimate in themselves but carried out in a manner that avoids taxation." At the Sarp border gate, the trade is made with a document, also used in the Türkgözü customs, called a "special invoice" that also stands for a declaration. This document ensures tax exemption; thus the commodities are bought for cheaper prices. This is not something that happens in every border gate. It is the ability to do trade with the special invoice that provides the opportunity for the suitcase trade in this region.

> Look, this was magnificent in the Eighties and Nineties. We, for example, had dealerships, and we still have. For example, I was the dealer of the region for Coca-Cola. The Georgian would come with their car and would say to me, "Give me 1,000 cases of Coke," they would say. In a single transaction, 1,000 cases, 2,000 cases. I have sold like that a lot, and then there were border trade cards, a special invoice for the border trade. This still exists in foreign countries. For example, I went to Italy last year

and bought a shoe from there. When you declare a thing in customs, you get back the tax added to the price, and the same deal used to exist here too. Coca-Cola used to give it for a little bit cheaper. There was a scarcity, because it went to foreign countries. Those do not exist anymore. Factories are set up there.

(ML_IW-19)

The suitcase trade became more frequent when the dynamism in the Russian markets started to wane, that is to say, when both those who came from the other side of the border did not have any stuff to sell left and when they achieved a certain amount of capital accumulation that could enable doing trade on a small scale. In essence, mutual trade started in this period. This time, Georgians started not only to sell the personal belongings they had at hand but also the products that they bought cheap in Turkey, both in their own countries and in the countries that are situated farther in the interior. As a result, the trade that is done, thanks to this border, influences not only the countries that border each other but also a broader region. This period was encouraged by the legal regulations of the two countries that made the trade easier (like the lowering of custom taxes).

Suitcase trade is a process done by the people who pass the border on foot. It is the name given to the process of their passing the border with the amount of goods that they can carry by hand and selling them and buying goods from Turkey (especially textiles). However, after a while, because of the troubles that were experienced during this period, which originated from the style of trade of the shopkeepers in Hopa, Georgians start to buy goods from other cities instead of Hopa. With the goal of earning a lot of money in a short while, low-quality goods were sold at high prices. Georgians who went to other cities came to realize this and stopped to buy goods in Hopa. The trade in Istanbul Laleli came about as a result of this process. Even the kids found ways to earn money in this dynamic period that had started with the Russian bazaars.

There had formed a source of income around the border without any cost. And what was that? They were bringing lots of furniture during those times from across, and they were carrying them with hand wheels, and with its worth at those times, $5 was a huge amount of money, not a small amount of money. Each turn of the kid was $5. They had at least ten turns a day. Can you imagine? Fifty dollars a day goes into the pocket of a kid, and what could that kid not do with that $50.

(FL_IW-29)

According to the data obtained from the Hopa customs accounting office in 1988, the total number of visitors entering and leaving Turkey from Hopa gate was 808. This number in four years increased to 1,400,120. No matter what the purpose of the trip, Hopa became a beaten track. Various types of transportation such as international transportation, taxi driver, or *dolmuş* driver (a dolmuş is a shared taxi or minibus running a predetermined route) became a major source of income, even for the children.

> The people who only had a taxi under their command and carried people during the period when the gate was first opened have today been able to start corporations and buy in to houses worth $300–400 billion without blinking an eye... Money was made; I mean, really, money was made quite nicely. But there was also this. They kicked their own bread. We would go down to the street and saw more or less the manner in which our people here would treat them. If a woman had a lot of bags, they would shout at her a lot. Though repeatedly cursing her, they would not take her to their car, because in any case there would come a person with fewer bags. After all the people were flooding, in the end with kicking and striking, in their words, they would reject them. Today they pray for those with lots of bags to come.
>
> (FL_IW-29)

In a system that was established in order to earn money, the thing that does not earn any money or that inhibits people from earning more money was being tried in order to get out of the system, like the women in the example above. To say it briefly, the wheels of capitalism started to turn in parallel in both of the two countries. For Yükseker (2007, 61) more than state-level interactions or the activities of global capital, it is the movement of goods, money, and images that creates social links between disparate societies that had minimal non-official contact with each other until about two decades ago. It is important to notice that the people involved in the shuttle trade are the "socially and spatially situated subjects" of transnational interaction rather than isolated units acted upon by an overarching process of globalization.

This process that meant entering the global economy was also a process that changed the structure of capital ownership on Hopa. People who had no capital earned money in a short time through easy means, and a lot of these monies were transferred to either the border trade or to the construction sector. The Hopa economy that was held up by those who were working in state-owned corporations and tea cultivation until the end of 1980s lost both its worker and peasant culture through time—at least a huge majority of it. Behind the success stories at the border lies, as mentioned above, the ability to be included in legal or illegal networks of relations. For the initial period in which the border opened, in the subtext of the story of "they made trillions in three months and then moved to Istanbul" told by the inteviewees at the border village was the emphasis on smuggling in large amounts, especially in the initial period. During these times when economic activities at the border were not yet systematized, when they had not yet gained an institutional way of working and Mafia-like relations had not yet been established, some people saw the gaps created by the system and created a significant wealth accumulation by involving their connections beyond the border. According to Roitman (2006), an accumulation regime that became autonomous from the state emerged at the border areas, and those who controlled this regime also facilitated the redistribution of the wealth. As such, when the border was opened, those who achieved wealth accumulation through smuggling—especially gold and uranium in large amounts—also developed mechanisms to sustain relations (with both official institutions and other people at the local level) that enabled the continuity of this wealth. However, in contrast to the data from a study that Şenoğuz (2014) conducted in Kilis in the Turkey-Syria border area, those who achieved this accumulation at the Sarp border migrated and settled in large cities. In the case of Kilis, what facilitated social acceptance of wealth accumulation created is that those who achieved the wealth accumulation shared a part of this wealth with the local people through social welfare and charity services that should have been supplied by the state or the municipality. On the other side of the Sarp border, the accumulation achieved, especially in the initial period, was transferred to Istanbul, and those who accumulated wealth migrated. I will elaborate on this point some more when addressing the issue of smuggling below.

Heyman's analysis (2014), in which he defined "unequal economic exchange between territories where values are differentially defined, such as unequal wages and social redistributions, but that also are connected across political borders," emphasizing the political economy of the border, has been quite instructive when addressing economic life in the initial period after the border was opened. Economic and political realities of the states on either side of the border was definitive in the mutual trade in the Sarp border area. Even looking at the products traded, the advantages brought about by the border for those on the Turkish side of the border and the distribution of capital present at the market in this period reveal this inequality. While larger accumulations of wealth emerge on the Turkish side, the border economy turned into a strategy of survival for Georgians. Immediately after reopening, the Georgians took all the goods for basic needs, and all consumer goods that were forbidden on the other side because of their definition as luxurious consumption goods, from Hopa and returned to their country to satisfy the demand.

> When saying trade in Hopa, it was bubblegum that was sold the most, bubblegum and Coke. Georgians bought cases and cases of bubblegum here. The wholesalers could not match the demand. Not like just one piece or two pieces, not with boxes, but with cases. What kind of lack that was, I could not understand. This is bubblegum, after all. It is not like a very important thing you see. Chocolates.
>
> (ML_IW-10)

Those who came from the other side of the border took as much as they could carry, on a very wide scale, with them and made sales by entering and leaving on a daily basis. After selling the goods that were in their hands, they bought both the goods for their own fundamental needs and the goods that were considered luxury products on the other side of the border, as we have stated above, and they returned to their countries and earned money by selling these goods there.

> From here all kinds of products like onions, potatoes, from clothes to food were out for sale. From there came copper, iron, a large amount of gas within the context of border trade, and timber... And in here, shops were so advanced. Clothing was so advanced, because can you imagine it? Eight thousand people were coming from the other side to here, and 8,000 people was such a huge crowd. There were enormous lines at the borders. Those people did not have a lot of money in their pockets, after all. They did not know Turkey. They crossed the border, did their shopping here, and then returned. They returned with great ease, so much so that sometimes they would return on the same day. They came in the morning, did the shopping, and returned in the evening. However, in that period, I was a baker at the Sarp border gate and sold 3,500 loaves of bread. I sold 3,500 only on the border, you see. It is so interesting that the sodas that were sold, Coca-Cola and biscuits, those kinds of things, I had a market, and we sold whole boxes filled with gum, 50 to 100 boxes of gum. They did not have bubblegum over there. They did not have any luxury consumption goods; any jeans, any gum, any leather, and absolutely not any fur.
>
> (ML_IW-19)

As the respondent above relays, trade of a broad variety of goods was done. However, while industrial goods were bought from the other side of the border, consumption goods were being sold to those who came from the other side of the border during the initial period. A similar unequal relation between developed and underdeveloped countries is being constructed at a regional scale, and goods of low value are being sold in high prices. The discourse that those who came from the other side are poor people and victims in fact was launched in opposition to the Socialist system and the disadvantages it created. However, the goods that people on the other side were depraved of were consumption goods like gum and chocolate that capitalism turned into needs.

With the opening of the border, everybody who lived in Hopa benefited either directly or indirectly. Those who did not do border trade or did not engage in the transportation business earned money from side sectors. The bakery sold bread at the border, and the market sold water or biscuits to those people crossing the border. It is easy to imagine the boom that 8,000 visitors per day could create, especially in such a small-scale economy as Hopa. Those who had nothing to do with the trade benefited from the cheap products that were sold in Russian markets.

For example, when they arrived, when the border was first opened, there were no hotels in Hopa to stay. A large proportion of the houses have turned into hotels. There were no places to stay for the folk, the guy said, so let me vacate my house and make a place for staying, he said. Here a lot of the houses were hotels. The police force, in order to arrange for places for them to stay, made the tea houses, restaurants, and places we call cafes stay open 24 hours out of necessity. I mean without forcing them.

(ML_IW-24)

After a while, traders who had nothing left at their disposal started to sell materials like iron, copper, and aluminum coming from dismantled factories and workshops. In fact, the majority of smuggling rested on these sorts of materials. Besides, legitimate corporations involved in this business surfaced in Hopa, albeit at a larger scale. Suddenly the number of corporations reached the hundreds, which caused a scarcity of office space in Hopa. This was a period in which all of the locations were transformed in a form that was suitable to trade. Not only the houses were turned into hotels, but the streets and offices were reorganized in accordance with this trade, and a seriously dynamic commercial life surfaced in this period. As it will be seen in the example below, people could carry a good that they sold to the other side of the border in their cars as if they had sold it to a person in their own district. The two sides of the border became "a single market."

The Georgian gate was opened in the year 1989. There was business to do; there was a lot of traffic. A market was established here by the Georgian side. There was a flashy trade, and through repeated crossings, the goods in Georgia exhausted. They couldn't bring anything new here, and when that happened, they started to buy goods from Turkey, but we couldn't satisfy their demand for some goods, especially clothing. All of Hopa's shop windows, shops, and markets were renovated. All the avenues and such were arranged accordingly, and money started to be made accordingly. The earnings were made through the scale with small amounts and a little bit of profit. We also sold electronic durables and junk from time to time. I even carried them with my own car to Batumi, which means that there was a huge amount of trade. That trade faded away with time. It has diminished a lot, especially in the last five years.

(ML_IW-18)

The commerce that was going on from the most fundamental food items to luxurious consumption goods, and on an increasing scale, started to decrease after a while. Border trade, in principle, included the satisfaction of the regional needs of the people who resided on the two sides of the border. According to Öztürk (2006, 109) the aim of this trade was ease in supplementation of the goods that people who live in the border regions need, contribution to increasing the context of mutual trust in the border regions, bringing dynamism to the regional economy, reduction to the minimum the smuggling of any kind of goods.

However, reaching these aims was closely related to how the relations between people, institutions and countries involved in this trade were built. At the same time, this trade was related not only to two states in two sides of the border but also to the global economy. In fact, the economy created at the Turkey-Georgia border region wasn't only between these two countries; trade with Azerbaijan, Kyrgyzstan, Uzbekistan, and Armenia was also made along this border. Macroeconomic systems tried to control the bazaars, and thus the government decisions about the border affected both the bazaars and other countries' share on the bazaars. Because of that, trade was interrupted from time to time. During the 1990s, according to Dursun (2007, 149),

> Georgian new market area had begun to be an answer to many of the structural problems–such as unemployment, lack of a market within easy reach, lack of investment and business innovation-especially in border areas, at least in the long run.

As the economic and social problems on the other side of the border started to be resolved, expectations from the border trade have increased too, and they started to produce themselves the goods that they had to accept in any condition at the beginning.

Is the cross-border trade settling down? Uncertainties and risks of border trade

Russian markets set up along the Black Sea coast from the Sarp border gate to Trabzon during the first several years began to dwindle in time as products for sale diminished. The local area was saturated with these products, and new opportunities emerged at the border. Those who were in suitcase trade moved to places where they could create a wider market, in particular to Istanbul. This time, in the Sarp border area, there was a transition to a new phase that forced those on the Turkish side of the border to mobilize and in which trade began to expand inland from the borders. In this new phase, the winners and losers of the border began to change, and a new set of actors began to be effective in economic life. This trade turned from commerce that individuals were conducting to a stage where corporate companies and capital owners invested and started to control the market in time. It is possible to define the trade that was done during the first period as a trade that was aiming to profit from scale with "small scale, low profit, high sale." However, as the potential of border trade was discovered, big companies originating from Ankara, Istanbul, and international corporations entered the market in Georgia, and this trade was acquired by big capital holders. People from Hopa who did not have the strength to enter this market or compete with big capital owners were left to mediate in this commerce and provide the transportation of these goods (that is to say, do international transportation). According to Donnan and Wilson (1999, 88) the border trade included three fundamental activities. These were prostitution, the passage of undocumented migrant labor, and smuggling. In addition to these, in our case, it is possible to consider two more income-generating activities common in border regions: the mediator-commissioner sector and international transportation. The border was seen as "the gate of opportunities" during this period and people were inclined toward transborder trade: selling all their possessions, borrowing money from their relatives, or bringing the material means of all the members in the family. While a lot of them were doing the business of mediating by opening up border commission companies, another segment of them, and especially the Hemshins, started the international transportation business.

> Do you know what the people of Hopa do? The people of Hopa know trade only as such, only as transportation. For example, a guy has a factory in Istanbul. He carries that guy's load, does commission in the border gate. There is no self-producing together with self-marketing. Always looking for ready-made. Bought a truck with bank credit, established a business, and put seven to eight trucks in that business. Together with that, he takes loads in Istanbul, takes them to Tbilisi, takes them to Batumi, takes them to Kazakhstan, Uzbekistan, Turkmenistan, even to Russia and to Ukraine with roll-on/roll-off ships from Zonguldak. They work toward that side from there. That is, we are good transporters. There is nothing else.
>
> (ML_IW-13)

Hemshins who held almost the entirety of the international transportation sector were in a more advantageous position on the other side of the border in comparison to Lazis. The thing that created this advantage has been both the fact that they did this job while carrying goods to Iran and Iraq and the network they had with the Hemshins and Armenians who were living on the other side of the border. The Hemshin who did trucking before the border was opened up put together money in their families and turned the trucks into lorries by taking credit from the banks. The Hemshini language, being a dialect of Armenian, provided another advantage in the cross-border trade. Alvarez and Collier (1994, 607) argue that Mexican truckers continually constitute and recreate ethnicity as part of an entrepreneurial process of successful penetration of foreign markets. They point out that "the ambiguities of identities in borderlands can also be strategically placed upon to forge, reformulate, and even mobilize ethnic identity to advantage" as in our case.

> Being Hemshin has provided an advantage for us. In any case, we did commerce with the people who were speaking the same language with us

on the other side. We went there with nonexistent capital, and we could bring back goods just because we knew Hemshin language. There was that kind of an advantage. We never experienced a disadvantage because we were Hemshin.

(MH_IW-22)

The economy at the border is a part of the global economy and the types of commerce that the globalization process has created. The people who enter this commerce are at the same time a part of the global economy. The transborder commerce of the Lazi and Hemshin is at the same time the entrance of these people into the global relations. However, even within the economies that are that big, the international trade has been continued through primary relationships: with the mate, friend, relative, and family bonds and their material support. During the period of no regulation at the time of the border's opening, a lot of people who did not have capital at hand found a chance to accumulate capital through extra economic factors (for example, the fact that they lived in Sarp, owned land at the coast, etc.). Those who acquired capital through exploiting the opportunities like extracting tribute, aiding smugglers, or smuggling themselves had chances to buy trucks and doing international transportation or international trade.

At least 80 percent of the Hemshins are not educated. Driving, mechanic—there is more employment through these. Naturally, Hopa is a place of transit, like transportation, etc. When it is like this, many of the employment and opportunities for jobs are created in these sectors. And naturally people earn a little bit of money. Again, many companies are established on this subject, and it is again those companies that earn the most of that money. What I mean is, for example, a driver who works with a lorry can only take 10 percent of what he makes. That means that if there was a driver, we would have a lorry. Today, a lorry is $200 billion. Especially among the Lazi, there is the idea that "most among the Hemshins own a lorry," but they don't know that 80 percent of that lorry is under debt.

(MH_IW-15)

Living in a place of transit created the most employment in the field of transportation. The means to earn money both as the owner of the job (the owner of the international transportation company) and as the worker (the truck driver) opened up. However, it was again the capital owner who gained this job, or those who put together capital with the help of their relatives became company owners.

As indicated by Baud and Schendel (1997, 220),

people on either side of the border may live in vastly different social and economic circumstances. Where income, employment, and life expectancy vary sharply, a border can mean the difference between poverty and material well-being and occasionally between life and death.

In the case of Hopa, though, the emphasis was more on the new relations created by the inequality based on the difference between the poor and the wealthy. Indeed, these new relations were precisely the effects of the existence of vastly different social and economic circumstances. For the people of the other side of the border, there was no choice but the one that entails buying and selling whatever was open and offered to them, at the price that was offered by the same token, on this side of the border in an environment of using the available resources and circumstances to their limits to make as much profit as possible. Yet this understanding of making the utmost profit using every means possible caused, in the long run, a significant reduction on the profits in comparison with the time where the border was first opened. The head of the Chamber of the Trade and Commerce (pers. comm.) summarized the period as follows:

> The Sarp border gate has opened in 88, and with its opening mass waves of people have started to come to Turkey from Soviet Union and other countries that were connected to it in such a way that there were times that 8–9 thousand people passed the Sarp border gate. Of course, the region was not ready for tourism in such a scale. Everybody occupied themselves with turning their houses to motels, this has brought in a decent income, because Hopa was the closest district, the people who have come from there have known here, and returned after they have done their shopping in here... Following this with the establishment of the border gate transportation with a large number of vehicles started, the mutual trades have started.

Border trade has been preferred over import-export, because border trade requires much less capital and experience; included goods are duty free or have lower customs; working with neighboring country reduces transportation expenses; deals are done easier with people across the border (Özçiloğlu and Sakar 2011, 23). Even though the potential of border trade was realized, both because of the lack of social capital and because of the problems of infrastructure, it was not possible to benefit from this trade enough. According to the president of the Chamber of Shopkeepers and Artisans,

> Our deficiency had been the fact that we did not know tourism well. We were not ready, how to treat a tourist how to speak with them. There is no language thing, and we don't know tourism and we looked down upon the people, if the object was 10 lire, we thought they are tourists anyway let it be 15 lire. They of course first came to Hopa and then slowly went to Rize and Trabzon and as they went to Istanbul they are a little bit knowledgeable for example when they are going to buy a pair of pants

they look and see some have brands and labels, from there a lot of them went by learning the business like that. They don't come here anymore like the old times when they are going to buy something they go to Istanbul or Izmir or some other place.

Rather than buying and selling products themselves, the people in Hopa take intermediary fees by making those who have products on the other side of the border and the companies that are in centers such as Istanbul and Ankara. Like Polese (2012, 22) says, "people matching a demand and supply on the two sides of the same border and earning money out of it." However, there are a lot of problems that have been experienced in this process that are not subject to sufficient legal supervision. Ordinary citizens on both sides of the border turn into international traders in a short period of time. This encounter with a different culture and language, not only of the neighboring country but also of commercial activity, brought a lot of difficulties with it. The hardships and risks of this new economic environment have been furthered by the rise of illegal means, tempting many who seek shortcuts to higher profits.

> I was doing a cellulose business. From out of the country, 6 meters tall, 80–90 centimeters thick as a log that was not cut through, I was doing those. And I was earning a good amount of money. This earthquake that hit Yalova and Duzce hit me too. I gave wares to the other side without invoices, and that person could not declare them... If I prepared an invoice, they would have declared that to the state, and I would have taken my money... I made a big mistake and as a result went bankrupt. It did not go out of pocket. It went out of profit. I did not have that much money in any case. I started with $1,050 dollars, but $125,000 has gone in a day.
>
> (ML_IW-6)

Border trade, unlike other commercial forms, is regulated by the mutual agreements between two countries. Border trade has been an incentive to develop the local economy for the cities and regions. In some cases, border trade has been useful for the country's economy, especially for the imported goods with low or no supply. For instance, in eastern and southeastern Turkey, diesel trade with a special license (Border Trade Document), given by the governors of the border towns, was distributed in other regions of the country. Such activities had impact on the national economy and displayed fluctuations, depending on the condition of the national markets.

Small-scale tradespeople and the ones who were involved in this trade became not only part of national economy but also part of global economy, and actors of this economy are exposed to risks at local, regional, and global levels. Today, there are many places of business run by people from Turkey on the other side of border. It ranges from small retailers to huge transportation and custom companies. Building these economic relations took many years, and for a long time, people had to trade without rules, in uncertain and highly risky environments. I should remark that it is a risky trade due to dependence of the border trade regulations and affairs of two different economies and states, and there exist uncertainties and unfamiliarity most of the time due to different cultures.

Those who were trying to do trade through the border between Turkey and Georgia faced very serious problems when the border trade first started. Both the people and the legal regulations were caught unprepared for the new phenomenon of border trade. The losses were also huge because of this. The border was a new border, and on the other side was a new country that had just declared its independence. Everything was new, and nothing had set into place. Despite the above mentioned high risk factors, border trade offered many otherwise unimaginable opportunities for the locals. With the lack of legal regulations, many risked their possessions and even their lives to enter into the border trade.

Traders involved in transborder trading practices may take a double risk regarding the ways they challenge the state (or state authorities) compared to those who take part in illegal markets in only one state. One source of risk derives from the illegality of markets. At the same time, risk derives from the practice of transborder transactions: People who in order to bring goods from one side of the border to the other hide goods and declare them incorrectly or use unofficial routes face the risk of being detected by border authorities. With regard to the specifics of transborder, small-scale trade in the broad field of informal economic activities, the risk of being detected by border officials always has to be weighed against the potential profit (Bruns and Miggelbrink 2012, 13). The uncertainties and risks that were created by an economic context when rules were not set during the time when the border was first opened were affected by the border policies of the two states, which kept changing incessantly.

Three agreements signed between Turkey and Georgia in 1992, the Trade and Economic Cooperation, Investment Incentive, International Highway Transportation Agreements, started the commercial ties between these two countries and were followed by several other agreements. Some of these agreements were abandoned either before or after taking effect. The Free Trade Agreement dated 2007 took effect in 2008, but it was revoked for the reasons explained below. The things experienced during and after the Kemalpasa incident is an example of the precariousness of the lives of the border people and their dependency on the economic policies of both countries.

The Kemalpasa incident: "That was an adventure that happened and ended"

Kemalpasa is a small town (*bucak*) that is linked to Hopa. It is between central Hopa and the Sarp border village with a population of approximately 4,500. Despite being ten minutes away from the center of Hopa, it is a comparatively calm place that the entertainment sector has not entered. However, shortly after Georgia and Turkey declared a tax treaty in 2008, Kemalpasa became a center of border trade. Hundreds of shops, generally of textiles, opened. Thousands of people started to come from the other side of the border every day, buying merchandise to be sold back in their country. The head of the Shopkeepers and Artisans Chamber (pers. comm.) describes the Kemalpasa Incident as such:

> It was during the period in 2009 when exactly the crisis took place. The crisis had started in the second half of 2008. You know what, let me tell you: we once again have entered the market late, believe it or not, there were not many local shopkeepers. They were from Denizli, Gaziantep, Rize and Trabzon, wherever you can think of from Anatolia. Many big companies started business. It increased from once a week to two, when they first came it was on Fridays and Tuesdays and the people who came weekly there were at least 6–7 thousand people who came every week. This commerce had started in 8:00 to 9:00 in the evening and would end at 3:00 to 4:00 the next day.

Kemalpasa, with its population of 4,500, turned into a trade center to which investors from all around the country came to sell their goods and thousands of people from the other side of the border come to shop. The people of Kemalpasa benefited from this potential that appeared suddenly by renting out their houses and barns and opening up places for the food and beverage needs of the visitors.

> The people of there came here and like on textiles or on different jobs, like house or kitchen gadgets... It had gone on dynamic like this for five or six months. They opened up shops, I would say even in the barns for their kids. They rented out the places that they would use as a woodshed, and that is for a lot of money too, but whatever was played, they ended here... The Turkish Republic was here, textile shops from all around, like how should I say, there were almost 500–600 shops that were opened. There are still ones that are open but now in worn-out condition... They have put a tax. An ordinary citizen would come and, for example, could leave by doing \$50–100 worth of shopping. That is, they could take five to ten pieces. This time when it became dense, 2,000 people started to come daily. If I had to make up a number, 1,000 or 1,500 people would do their daily shopping and leave.
>
> (MH_IW-11)

This trade that started with the removal of the customs tax, caused, similar to what was experienced in Hopa during the time when the border was first opened, the rearrangement of space in Kemalpasa too in accordance with this commerce. The locations changed, the daily life was organized according to this, and Kemalpasa turned in to a huge market. Those who came from the other side of the border came with ease as if "the border did not exist," satisfied their daily needs, and returned. This event essentially shows how the borders set for political reasons and used as a "barrier" can be discarded first by the states and then by the people when economic interests are at stake.

This exchange created new opportunities as well as aggravation on the other side of the border. Georgians were making profit by selling the merchandise they brought from Turkey at lower prices than the local markets, which eventually created the local shopkeepers' legitimate grievance. In the words of one of our Georgian respondents,

> Georgia's economic condition had deteriorated a lot. Very cheap goods came to Kemalpasa. Because it was sold cheap, they also sold cheap to the other side. Goods that came from other places, from Iran, from Iraq, did not sell (in Georgia). A lot went from here now, because the economy had deteriorated a lot. There the cheaper of goods sold more. The goods that went from Turkey were sold. But whomever it was that came there earlier and set up a business, whether it be Italy or France, there are a lot of goods that come from Iran in Georgia. Their economies deteriorated. They closed it because of that, you see.
>
> (FL_IW-9)

In the article in the *Milliyet Newspaper* titled "Laleli Dream of Sarp Has Ended Early" from 20 April 2011, the words of one of the shopkeepers of Kemalpasa were quoted: "There were certain agreements between two countries based on the suitcase trade that were valid until the year 2017. We have come here trusting that. But Georgia has unilaterally declared this agreement null." Another shopkeeper's elaboration in another news article is as such:

> Those who come from Georgia cannot pass the goods that they bought from the gate. A high tax is asked of them. For this reason, 70 percent of the shops are closed. We had two shops, and we closed one of them and will be closing the other one too.

As the president of the Shopkeepers and Artisans Chamber (pers. comm.) asserts,

> This time the shopkeepers from there had risen up now logically. I as the head of chamber of shopkeepers would stand up too if something like that happened. What you are doing, we are paying taxes to you here. It was exactly as they say, it was itself exactly a factory without chimneys. I mean that business was suddenly cut off. However, there was another side to this business. When this new border gate building was built, the presidents of the two sides, the president of our side and their president, said, "Five years without interruptions, and these businesses will again continue with

certain subsidies after five years. They were saying, "What are these gates? Let's open these gates. Let's make trade with pedestrians on the seashore with a walking system. Let's make this, let's make that." These people weighed these businesses too, but of course these promises were not kept.

An event, which appeared with an agreement that was signed by the political authorities of two countries and created a source of income for the citizens, suddenly caused the citizens to turn into victims of these political decisions by disregarding the potential losses that they would have. It can be seen here that not only the economic interests of the two countries were at hand but regional and even global economic interest relations came into play. Georgia has not only been a market for Turkey but also for Middle Eastern countries like Iran and Iraq and European countries such as France and Italy, and it has been a market for America too. As a result, it was not possible for the two countries to enact the rules of a global market on their own, and the agreement that they signed was rendered void. Border people, who were between political decisions of two states, had to accept risks and uncertainties in order to continue their lives and to improve existing conditions. One of our interviewees who identify himself as a "border victim" remembered his experience of running a shop in Kemalpasa during the border trade boom:

> People from all around the Turkey were coming, opening up shops and earning money. We already knew about the textile business. We had pondered on it and thought, "The guy comes from such a long distance and makes money by opening up a shop, so why shouldn't we?" This was already our job... We opened up the shops, and there was an enormous amount of business. We were earning like $4,000, $5,000 a night. When he earned $3,000, I was saying, "Is it that low?" The business was so much that my husband couldn't find enough bags. He hired two Georgian workers. The woman used to be a doctor, but she was working with us. She said, "I make better wages here." This lasted for three to five months... there was no tax, the Georgians were not paying tax for the goods that they bought. Then they started to apply taxes at the border like a bang... you had to pay taxes according to the weight of the goods you bought. You would pay at least 600 lire even if you did shopping worth 300 lire. The business ended with a bang... everyone crashed and burned, as we bought the goods with credit, you know. We had given checks for three to five months... the business ended before the due dates, and we were stuck with the goods on hand.
>
> (FL_IW-31)

As Kolossov (2005, 632) asserts,

> for large enterprises and especially for transnational companies, customs formalities and taxes rarely play a significant role, while for small and

medium enterprises located in border areas, they become a serious obstacle stimulating them to re-orient their activity to the domestic or local market.

These economic disturbances affected the small- and medium-scale enterprise owners the most. Today, a citizen who crosses to the other side has to pay a tax even for something they bought as a gift.

> In Georgia, nobody comes anymore, because nothing passes the border. Now I can transport the stuff that belongs to me when I am going to my mother. If I buy a gift for my mother, they have put a restriction in place. You cannot buy anything.
>
> (FL_IW-9)

This dynamism in trade that lasted around six months ended with Georgia starting to excise duties on border trade. Today in Kemalpasa, it is possible to see hundreds of shops whose shutters are closed. In a similar situation in Benin-Nigeria as Flynn (1997, 312) explains, border residents have responded to the decreased trade traffic, omnipresent custom guards, and plunging economic opportunities by forging a collective "border identity" based on their territorial claims to the region and their perceived right to participate in, and profit from, transborder trade.

But in our case, this kind of collective identity has not emerged most probably because of the existence of different ethnic groups and of people from other regions of the country. The closed shops and desolated streets of Kemalpasa remained as the signs of aggrieved shop owners whose business venture tumbled because of a decision-making mechanism beyond their reach. A decision taken by Georgia directly affected the lives of the people on the other side of the border. According to Baud and Schendel (1997, 220),

> Border economies are always strongly influenced by political measures, and political processes on either side of the border do not normally coincide. Border economies react instantly to short-term policy changes, and constant adaptation lends them a speculative, restive character. This is one reason why it is so important to treat the region on both sides of the border as a single unit: changing economic policies on one side of the border lead to immediate adaptations on the other side as well.

However, although both sides of the border quickly adapt to these changes, the risks of cross-border trade increase in each transformation period. These risks are material and at times reach life-threatening dimensions. Those who are involved in border trade in the Sarp border area have faced risks posed by political uncertainties rather than dangers of the illegality of "illegal" trade, which we will discuss in detail in the section on smuggling in this chapter. On top of the risks that appear as a result of legal regulations, inadequacies that result from local authorities, or people in the trade that is conducted on this side of the border, similar risks that exist on the other side of the border are added in the transborder trade, which require larger capital, and the financial risks are doubled. Of course financial risks are not unique to border trade. Many economic activities expose people and businesses to serious financial risk. However, our case includes a unique type of risk factor that could be identified by the lack of regulations. As the president of the Chamber of Commerce, who himself is a commission merchant, says,

> The firm has sent the goods, but how is it going to send the earnings from them? How is it going to collect the earnings they make? They will send the money up front and then think, "Will the Turkish firm will send me my goods?" When you send goods, is the company in Georgia going to pay you for these, and if so, how will it be? You get your money... there were people who would intercede with you on the street who are dressed up as police who would search you and take your money that you carried with you. That is to say the commerce was done in great conditions, but as I said, in the period that is close to the last ten years, this does not take place anymore.
> (President of the Chamber of Commerce, pers. comm.)

Both the potentials of the trade not being understood and the rules of trade not being established on both side (and the human factor) has caused the financial risks and losses to increase. According to Dursun (2007, 177), the economic relation between Turkey and Georgia is an "asymmetric" one. Turkey constitutes a model for Georgia, which is undeveloped with respect to Turkey. Especially in the first years of independence, Turkey was the only door for Georgia to reach the whole world. Georgia gained importance within the system of global relations due to its strategic location. The Rose Revolution in 2004 clearly put forward this significance, as Georgia became a good market for many European countries and America just after this revolution. However, Turkey could not be prepared for such a development and could not realize necessary legal and infrastructural projects. As a result of Turkey's belated move to use the potential of Georgia and increasing competition in the Georgian markets, Turkey lost her market. The irregularities between the two states affected trade relations at various levels, and companies started to lose their commercial ties with Georgia. Lower-quality goods that brought large profits started to lose their place in the markets.

> Now what was done in Turkey? They started producing cheaper and lower-quality goods. They lowered the prices each time, but they lowered the quality too. My brother didn't lower the price but didn't lower the quality either. After all, the people on the other side are cultured. They know this, and they are aware of what happened after a while. Our

products started to be disliked. They thought that we produced lower quality, and that is what happened as a matter of fact.

(ML_IW-19)

This lack of regulation was not only experienced in Turkey but on the other side too, and it affected the trade that was done during the initial period a lot. While one factor in the emergence and increase of these risks is the unsettled condition of the system, another was the irregularities that people created.

We couldn't make communication. We did trade, but we slipped up. We paid in exchange for goods, and we couldn't get our money back. We couldn't find the guy, even... we couldn't go there once around 1995 or so. There was civil war. In Georgia, we couldn't go in those periods. We came face to face with police, the people, Mafia if I have to make up an example. We could hardly pass a distance of 500 kilometers in three days and that was as a group and with great fear. Whether it was transportation or another job, we couldn't stay or spend the night without a reference. Now, however it is better.

(MH_IW-11)

The attitudes of people in constituting cross-border relations—both economically and socially—attached much importance and were constituted by not only the historical commonness but also by the existing socioeconomic structure (Dursun, 2007, 161). Even despite the fact that there were people who had the same ethnic identity who lived on both sides, they faced cultural differences and the difficulties that this created.

They seemed were alien to us. That is, we couldn't get along with the Georgians. There were problems. Those who did trade were in trouble, that is, they started to pursue taking 5 lire when you gave them 3 lire. I'm saying it was just like that. What they did was stealing and cheating. In a sense, there wasn't commerce... You cannot trust. You see, they don't think in long term. In fact, when you reconsider we are neighbors, our cultural qualities are almost close to each other. They think in short terms, I mean, like "Let me con this person in a month for $5,000."

(MH_IW-11)

This new chaotic setting made the people of both countries, who lived side by side but did not have any relation for many years, vulnerable to various types of abuses and exploitation. People from Hopa, who sold the bad goods in their hands at high prices, experienced similar incidents when they crossed to the other side. Such cases must be seen as a consequence of the precarious economic conditions pushing individuals into illegitimate and often unethical means of earning money. It is a context where people struggled to survive and reach a level of economic security in the middle of transition to a new system.

It is said that there are close to 800 lorries that are registered today in Hopa. However, the people who sold their trucks and bought lorries to do international trade say that they in fact lost money because of the credit that they took from the bank and the interest.

> What contribution did it have to Hopa? I don't think so. It does not have any contribution to Hopa. Let me say something like this. Perhaps we earn $2,000 or $3,000 with the lorry. However, you buy a lorry for $240 billion, aged zero. While doing that, you are going to sell it ten years later, and you will sell it for $100 billion... it is the banks who are the real winners here. It is the banks who earn the most money and who don't take any of the risks either. The banks earn good money. In tea cultivation, the banks and Mercedes earn money first. Then the capitalist earns money if something is left. If a bone is left for us, we make do with that bone. The system is as such, you see.
>
> (MH_IW-2)

Alongside these material risks and conditions, what was more important was the nonexistence of security of life. People who disappeared or were incarcerated were frequently mentioned in the interviews. People's expectations were very high when the border was opened. The people were occupied with the goal of earning a lot of money in a short period of time. The same expectations were cited by Dursun (2007, 162) for the other side of the border. People expected that the transition would increase affluence. They also believed the state would provide full employment, free education, and health care and prevent the emergence of huge economic inequalities. But there was a remarkable decline in the status of professional, scientific, and technical occupations. Access to wealth depended rather on contacts with Western capital, connections with officials and elites, and involvement in criminal activities involving extortion, drug trafficking, prostitution, etc. (Dudwick *et al.* 2002; cited in Dursun 2007, 162–63) (Dunford 1998).

The irregularity on this side of the border is more intense than the other side in a country that recently declared its independence and is still struggling to become economically viable. On top of these, there also appears to be a problem of security. The people who take goods to the other side or bring them back try to move as a convoy without breaking up from each other. However, this cannot prevent robberies or swindling. This situation creates an obstacle for the commerce to be done within rules and causes many people to lose money. However, even more important than money is that people fear for their lives, because the cases of robbery and assault are frequent.

> Like most trade routes, it is dangerous. The danger is worst on the journey home, as the merchants return across the frontier with bales of Turkish leather jackets and cheap computers and bundles of greasy Western banknotes. Near Kabuleti, a few miles into Georgia bands of armed

robbers in military uniform ambush the bus convoys and strips the passengers of their treasure.

(Ascherson 1995, 196)

We went in fear. What will happen to us? Wen even the smallest thing happens to you, they say jail or $50,000 or $100,000. These are experienced even today.

(MH_IW-11)

The things that one of our interviewees described (his father was lost in 1994 on the other side, and he never learned what happened to him) display the hardships of doing cross-border trade in that period.

He was a driver, and we had a bus in the past. When this border gate was opened, trans-border trade started. It was around 1989 when it opened, and we lost our father in the year 1994 on that side, in Georgia… That is, he is still lost after 17 years. We investigated a lot for five to six years, on that side and on this side. We didn't reach any conclusions, and we are waiting just like this… He was going to go to Ukraine with a plane in the morning, beside the other products. Since that day, we haven't heard any news. That means that we have lost contact one day.

(MH_IW-11)

In her unpublished thesis about cross-border cooperation between Turkey and Georgia, Dursun (2007, 182) analyzes the experiences of Turkish enterprises in Georgian market. She states that "to cope with the turmoil in Georgian markets has turned out to be extremely challenging for Turkish entrepreneurs. They have complaints about the issues as legislation, unstable taxation, corruption, barter trade, and lack of reliable partners in Georgia."

In addition to those who are involved in cross-border trade in some way and those who face material risks created by this trade, the experiences of those who are often involuntarily pulled into these networks of relations, especially women who are forced to become sex workers, should be addressed. The entertainment sector established over the bodies of these women, although diminished at present, has been the sector that sustained Hopa's economy for many years.

Entertainment sector

The processes that I have mentioned above were not experienced in the form of successive stages following the opening up of the Sarp border gate. They were woven together and often fed each other. The appearance of the entertainment sector in Hopa was in a form parallel to the starting of the suitcase trade. After the opening up of the borders in the countries that declared their independence after the Soviets, things that were similar to what happened at the Turkey-Georgia border were experienced, and many people searched for means to earn money and continue their lives by crossing the borders. This process in which the countries attempted to adapt to the global economy at full speed has brought inequalities among the citizens of these countries too. Those who have been crossing the border have been people from the lower classes of these countries. What was experienced in Hopa has not been different from this. As Dursun (2007, 1983) states:

> In the first years of independence so many travelers had come on ships, not only from Georgia but also more from Russia and other post-Soviet Countries. All of the shuttle trade has run in these bazaars, and thus they had resided near the bazaar, and harbors. The number of the hotels in these areas had rapidly increased constituting a residential area for these shuttle traders and also for prostitution.

The entertainment sector includes restaurants, disco bars, hotels, and prostitution. This sector has three main economic agents: the owners of the places, the women who come from outside for sex work, and the men who demand this type of entertainment. This sector is believed to hold the economic life of Hopa on its feet today. This sector developed gradually after the border was opened. At first, dilapidated or abandoned houses or cottages in the tea fields were the loci of prostitution.

Like many of the interviewees have told, today what keeps the economy of Hopa alive is the entertainment sector.

> You see, there are six to seven four-starred or five-starred hotels in a town that has a population of 15,000 to 20,000. They have 70–80 employees working in each of them depending on the size. These are of economic value, but they are bad morally.
>
> (ML_IW-10)

The economic benefit of the sector and the employment opportunities that it has created in the district enabled both its legitimation and its spread. Nobody even knows the number of small hotels, bars, and discos. This sector provides at the same time employment for the youth in Hopa and makes the enterprise owners earn money as well as letting the shopkeepers in the environment earn money from this business.

> They would need a hair dresser, and they would go to the beauty salon. They would need clothes, and they would go to the shop. Their food, banks, hotels, discos, all of these have happened as links of a chain with them. You see, with us nothing works if the foreign women are not here. Restaurants don't happen, the hotels don't work, discos and the like. For that reason, this realm affects everyone too. I think there are 15–16 beauty shops.
>
> (FL_IW-25)

As the quotation above shows, the entertainment sector creates other job opportunities or feeds the sectors that exist. The shopkeeper who started to earn money over the opportunities of the entertainment sector does the trade he does at his location. The goods that he has are being redesigned according to this. Although many interviewees admitted that this sector is the major source of income for many inhabitants of Hopa, especially our female interviewees constantly voiced their disapproval of it. In the words of one of our female interviewees,

> We didn't know what a beauty shop was. There was only one beauty shop called Muazzez. Now there are many beauty shops opened up only for Russians. Disco is in Hopa, prostitution is in Hopa, beauty shops are in Hopa… Say, if you say, is there more of the other dining or useful shops or more of these businesses, these outweigh. For example, I am going to buy a dress suitable for me, there isn't one, you see… There are only ones suitable for them. If you go into a shop or go into a shoe store, there is never anything suitable for you, you see.
>
> (FL_IW-5)

While the streets, shops, and goods that are sold are organized in order to satisfy the needs that the entertainment sector creates, the living spaces of those who live in Hopa are discounted, and some streets have become unusable for the women. They had to forego the right to life for the entertainment sector that is being legitimized on the one hand as the source of sustenance for the household and the family. What almost every respondent stated was the fact that the biggest problem in Hopa is unemployment. For them, there are almost no job opportunities for the young other than the entertainment sector at hotels and discos or bars.

> This has to be accepted that there are things that have created a lot of employment as a result of this entertainment sector. There is no other employment for the young who work at the hotels anyway. There is nothing that is done. There were 550 people when I started working in the tea factory, but now there are 250. It is diminishing. There is not an increase there. The young work in the hotels. That is not commerce. They are working, after all, but this is a resource that is brought by the gate. And also the gambling—they gamble in the hotels. There are also the opportunities that are brought by that. There is also internal tourism, you see. A lot of people come for entertainment from the outside, especially from Erzurum.
>
> (ML_IW-10)

A major part of the demand in this sector that the border has created comes from those who come from the provinces who are in close proximity of Hopa. The sex tourism that the respondents call "internal tourism" also indicates that the zone of influence of the border has expanded. It has a widening effect from the position that the gate has outwards even though its intensity decreases. As an expected consequence, the entertainment sector based on the sex industry has caused the formation of criminal organizations. Crimes committed by Mafia-like organizations are a major source of disturbance for the inhabitants of the city. According to our interviewees,

> A Mafia has also been born here, of course, over this huge amount of money. What's more is those who do it are locals. There have also been people who started to take tributes when there was a lot of money made for nothing
>
> (ML_IW-10)

It can be understood that the entertainment sector has created other sectors and led to the emergence of new power relations from the above quotation and the gun fights that were experienced in recent times.

> The unjustness of this sector, those people who have gained more money from the Mafia-like sector or those who hold the power have come to places like a hotel, restaurant, or whatever. These are not things that normal people can attain. They have accumulated that capital, but that capital after all does not belong to Hopa. The extensions of bigger corporations or the entertainment sector did not use it as a contribution to the people or unemployment either.
>
> (MH_IW-15)

Not only the clients of the entertainment industry but also the employees, namely the sex workers from post-Soviet countries provided economic benefits and boost to the local economy, especially for the small business owners.

> They go to beauty parlors or they go to other places, and they do decent shopping, When they are going to their homes, they do their wholesale and take off. They never bargain either. They do their shopping pretty decently, and they provide a somewhat huge or small aid to all of the shopkeepers. In the smallest type of pressure, for example, it is heard immediately. None of them get out. They become afraid, because it is a small place. Being deported is not a simple thing like that.
>
> (The head of the Shopkeepers and Artisans Chamber, pers. comm.)

As a result, a structure has emerged in which capital owners keep operating with high profits. They establish the rules for, on the one hand, the small-scale company owners who are trying to benefit from the dynamism that this sector creates on the other. However, it is not possible to understand border economy as a whole without addressing the illegal or informal networks of relations that directly or indirectly participate in the emerging of this structure and facilitate wealth accumulation through illegal operations.

Illegal trade

If there is a border, there is smuggling too. Smuggling is unique to the border regions and is one of the fundamental bases of the economic life in these regions. Smuggling is the indispensable truth of these regions whether it is done by organized crime groups or by the ordinary citizens living in the border region. Scholars have used a variety of definitions in their studies of the informal economy alternatively called the "second," "black market," "unofficial," or "unrecorded" economy done by border people. "State boundaries have always been, and often still are, shaped by trade and a range of illicit activities that constitute important livelihood strategies for the border populations" (Mwanabiningo and Doevenspeck 2012, 85). The content and intensity of trade or illegal activities shape the security and permeability of economic and political relations of states on both sides of the border. The Sarp border gate was closed for 51 years as a border of two ideological and economic systems. Inhabitants of this region couldn't point at the other side of the border. The impassable Sarp border gate was full of barbed wire, watchtowers, and tracking fields, but on the other hand, the border became a source of income for the same people.

> One of the most interesting forms of creativity on the border involves smuggling, an illegal activity that borderlanders easily rationalize. To be sure, that is dishonor in smuggling strictly forbidden and harmful substances such as drugs, but not in smuggling ordinary consumer goods.
> (Martinez 1998, 313)

This situation is one that one of our respondents rationalized by saying,

> If the locals go and buy a box of cigarettes or buy a couple of bottles of alcohol, or buy meat, of course it is only a natural thing that the people who live in the border region benefit from this.
> (ML_IW-19)

This is the commonly held belief of almost everyone who lives in Hopa.

The initial period of a border's establishment is a period when smuggling (especially the smuggling of radioactive junk, drug trafficking, and prostitution) suddenly becomes abundant. The economic activity in this period develops in a manner more reliant on informal friendship and buddy relations rather than relying on the force of capital and hence in a manner in which illegal activity is dominant. James Scott, who read of the rebellion of peasants in Southeast Asia through the concept of "moral economy" as the right to subsistence against the attack of the market (1976; cited in Şenoğuz 2014, 111) contributed the concept to border studies literature. This model has been used to describe local economies that resist "the shadow sector that emerged during the transition period from the socialist market that disintegrated with the dissolution of the Soviet Union to capitalist market" (Hann and Hann 1998) and the transnational market created by multinational agreements by smuggling (Galemba 2012a; 2012b; cited in Şenoğuz 2014, 111). Border people who became a part of transnational market with the existence of the border often turned to illegal economic activities while trying to survive against these types of relations, on the one hand, and to adapt to political maneuvers of the state and regional and local powers on the other. What was to be smuggled was generally determined by the difference in prices or by someone in the state. For example, the diesel trade was let free with a decision taken by Turkey, and the people found ways to benefit from it more. Because one needed a border trade document in order to be able to do diesel trade, many people filed for this document in their friends and acquaintances's name and did diesel trade.

> It was during the years 1995–96… During that diesel period, we had many things. We did our permits in the streets, like $200 to $300 dollars for everyone. We gave to our friends and made them liable and got border trade documents. They would reap at least 500 lire each for themselves from us… This was allowed for a year.
>
> (Chamber of Shopkeepers and Artisans)

According to the increasing tax losses of Turkey from petroleum trade and also the increasing pressures of businessmen related to liquid oil trade, governments limited the range of goods that were subjects of border trades with a decision of the Council of Ministers on 22 June 1998 (Dursun 2007, 134). The diesel smuggling started after that.

> First, from here things like these companies were established, and pulling diesel was official, but now it's through smuggling. How so through smuggling? Let's say we have a car, you go and extend its tank, you go and fill it up, come back, and empty it and sell it. You go and come back again, but they don't have a chance to measure the diesel in your tank, or you use that if they control it. That has developed a lot. There have been people who bought a car just to do smuggling, but there were a few people who were caught too. However, I think it was a little bit to show off, because everybody knows this in here. Even the police know it, and the major knows it too. There is no way something this common can be hidden, you know.
>
> (FH_IW-3)

A lot of people who live in the region have also crossed the border, bought diesel, and returned in order to satisfy their own diesel needs.

That is also a source of income in a way. Let's say the guy departs from here to there empty when they get their own diesel. For example, a taxi driver uses that gasoline for 15 days in here. Let's say the guy has a pickup, and there is a person who commutes from their home. They have a private car. When they nicely fill up the tank, it would last for a moth. That is, you see, a contribution to their own budget.

(ML_IW-24)

Beneath the perception of smuggling as an extra income source/subsistence strategy that "contributes to the family budget" lies the belief in rightfulness created by being "condemned" to live on the border. Border people have to create a living, knowing that every political decision and regulation by the state they live on the edge of will rapidly change their living conditions.

As I indicated before, any tax regulation in both countries, and especially the price differences between the goods, immediately affects the trade in these regions and smuggling activities. The goods that are smuggled the most in recent times in Hopa are cigarettes, diesel, meat, honey, and alcohol. Most of the people perceive this as an income that contributes to the family budget and an income that is the right of the people who live around the borders.

For example, our friends have certain cars, and they at least go to buy diesel or go to buy sugar. They can bring a couple of items when they are returning. I have a nephew, and he had a taxi. He crossed once a week and bought his gas, bought his diesel. Surely there is not like a great differential, but you see, if he saves 100 lire from gas, if he brought five kilos of sugar and he saves 20 lire from that, for each tour of going in there and returning, he benefits by saving 120–130 however you may look at it.

(ML_IW-8)

However, there are shop owners who are not pleased with this business. Especially the shop owners who sell the commonly smuggled commodities express grievance that their business is hurt because of smuggling. A shop owner who runs a cigarette distribution agent complains from the lack of controls and the implicit consent given to these activities.

What is it? It is below the counter. They cannot open up that counter. The police or the official from the tax bureau only looks here and asks if there is any that are smuggled. You see, of course, they will say there aren't any. Would they say there are some that are smuggled? This is the system of control. You see, for example, they get caught at the gate, and you say, "Why did you let him?" He says "Come on, I can't check all the cars, you know. I can't check 2 out of 15, I check randomly." However, they know their man, and that man makes an express transit. They ask why you have not checked them. He says I couldn't check all of them. He just happened to be among those who I couldn't check.

(ML_IW-19)

As Martinez argues, the types of smuggling that are small scale, unorganized, and done especially in order to satisfy daily needs are seen as a right of the people who are living around the border. They are not thought of as illegal activity and are not even labelled as smuggling (Martinez 1998, 313). Many transnational movements of people, commodities, and ideas are illegal because they defy the norms and rules of formal political authority, as Schendel and Abraham (2005, 5) put it, "but they are quite acceptable, 'licit,' in the eyes of participants in these transactions and flows." For the people of Hopa, the case when they are generally labelled as smuggling and not approved of in any way is the smuggling of narcotic materials such as marijuana, heroin, and pills. The people who are engaged in drug trafficking are generally blamed as being amoral and earning unjust income.

Thirty percent of the marijuana and heroin that enters Turkey enters from this gate. While entering from here, there are some spill overs. That means this business takes off from here. It is directed from here.

(MH_IW-15)

However, as the head of the public security office recounts, there is no trouble related with narcotics in Hopa. Marijuana, heroin, and pills are more expensive on the other side of the border, and that's why there is nothing that is coming from the other side. He said, "It must go from here, but there are no large scale busts here either," and he added that transporting through this route to Istanbul was attempted last year and was busted. In the following minutes of the interview, he said that cases related to narcotics were inevitable where there is an entertainment sector, but there is not a huge problem in this sense in Hopa. The head of the public security office said intercountry smuggling cases that result from price differences between countries can be solved through precautions and not through apprehensions. He mentioned that there is smuggling done at the Sarp border gate for a wide variety of things like cigarettes, diesel, watches, walnuts, and honey.

It won't be false to think of/define smuggling in three different densities. The first is the one done by the people who are living at the border and which they do more in order to satisfy their daily needs. This one that can be defined as a contribution to a home economy. The second one is again of the people who live in the border region, one that is again realized in a small scale, however the dimension can be defined as small-scale trade in order to sell the goods that they have at hand or sell the goods that they buy at higher prices by taking them to their own countries. The third is the one realized by organized groups done in large scale. The dimension of it especially includes smuggling of drugs and human trafficking. Here it should be noted that smuggling emerges in different forms and intensity on each border. In Hopa, smuggling is done today in a way that contributes to the household economy rather than in way that enables sustained high wealth accumulation. This way of smuggling is described by the interviewees on the basis of certain categories such as "legal" or "illegal" or "legitimate" or "illegitimate" instead of being defined as completely good or bad. As might be expected, smuggling of drugs or weapons was perceived as illegal or illegitimate and qualified as "immoral," and smuggling of gasoline, cigarettes, alcohol or foods was perceived as "normal" and even as a "right."

Smuggling in any case, whether it is the smuggling on a large scale done by organized crime groups or at a small scale (which can be considered a subsistence level done by the people living near the border on a day-to-day basis), is a situation that threatens the safety of the lives of the people who do it. Beyond smuggling, the border trade that is done legally also contains within itself the risks unique to the border.

Since the opening up of border, economic life that is connected to the border in its entirety has been established in Hopa, rules were determined on both of the sides, the people have gotten to know each other, and the "economic limitation" and/or "unlimitedness" of the border has come to be realized. For this reason, there is a commerce that is more stable and that is done by people who know what they are doing and what they will encounter. As Donnan and Wilson (1999, 107) state, "When people cross state borders, whether it be political or economic refugees, or as tourists, or to purchase or trade goods, they become part of new systems of value. These new value systems are simultaneously materialist and idealist." This new materialist and idealist system determines the social, cultural, and political systems of border regions as we will see in the following parts of the study.

Note

1 When addressing the issue of culture in border areas, Heyman (2014) focuses on the question of how culture production bypasses state control over interactions between two territorial spaces. When, on the other hand, it takes place in a space organized under state control, and he accepts this question as an important issue in border studies.

References

Alvarez, R. R., & Collier, G. A. 1994. "The Long Haul in Mexican Trucking: Traversing the Borderlands of the North and South." *American Ethnologist* 21, 606–627.

Ascherson, N. 1995. *Black Sea*. New York: Hill & Wang.

Baud, M., & Schendel, W. 1997. "Toward a Comparative History of Borderland." *Journal of World History* 8 (2), 211–242.

Bennafla, K. 2014. "Sınır'ı Tartışmak: Yuvarlak Masa Söyleşisi." *Toplum ve Bilim* 131, 11–32.

Bruns, B., & Miggelbrink, J. 2012. "Introduction." In *Subverting Borders: Doing Research on Smuggling and Small-Scale Trade*, edited by B. Bruns & J. Miggelbrink. Berlin: VS Verlag.

Donnan, H., & Wilson, M. T. 1999. *Borders: Frontiers of Identity, Nation and State*. New York: Berg.

Dudwick, N. 2002. "No Guests at Our Table: Social Fragmentation in Georgia." In *When Things Fall Apart: Qualitative Studies of Poverty in the Former Soviet Union*, edited by N. Dudwick, E. Gomart, & A. Marc with K. Kuehnast. Washington, DC: The World Bank.

Dursun, D. 2007. "Cross-Border Co-operation as a Tool to Enhance Regional Development: The Case of Hopa-Batumi Region" (unpublished thesis). Middle East Technical University.

Flynn, D. K. 1997. "We Are the Border: Identity, Exchange, and the State along the Benin- Nigeria Border." *American Ethnologist* 24 (2), 311–330.

Galemba, R.B. 2012a. Taking Contraband Seriously: Practicing "Legitimate Work" at the Mexico-Guatemala Border. *Anthropology of Work Review* 33(1): 3–14.

Galemba, R. B. 2012b. "Corn is food, not contraband: The right to 'free trade' at the Mexico-Guatemala border." *American Ethnologist* 39(4): 716–734.

Green, S. 2014. "Sınır Araştırmaları: Alanla Ilgili Bazı Düşünceler." *Toplum ve Bilim* 131, 32–44.

Hann, C., & Hann, I. 1998. "Markets, Morality and Modernity in North-East Turkey." In *Border Identities: Nation and State at International Frontiers*, edited by T. M. Wilson & H. Donnan. Cambridge: Cambridge University Press.

Heyman, J. 2014. "Sınır'ı Tartışmak: Yuvarlak Masa Söyleşisi." *Toplum ve Bilim* 131, 11–32.

Kolossov, V. 2005. "Border Studies: Changing Perspectives and Theoretical Approaches." *Geopolitics* 10, 606–632.

Martinez, J. O. (1998). *Border People: Life and Society in the U.S.-Mexico Borderlands*. Tucson: University of Arizona Press.

Milliyet Newspaper. 2011. Sarp'ın Laleli rüyası erken bitti (Laleli Dream of Sarp Has Ended Early). 20 April 2011.

Mwanabiningo, N. M., & Doevenspeck, M. 2012. "Navigating Uncertainty: Observations from the Congo-Rwanda Border." In *Subverting Borders: Doing Research on Smuggling and Small-Scale Trade*, edited by B. Bruns & J. Miggelbrink. Berlin: VS Verlag.

Özçiloğlu, M., & Sakar, M. 2011. Sınır Ticareti Uygulaması Sorunlar Ve Çözüm Önerileri. Girişimcilik ve Kalkınma Dergisi (6:1).

Öztürk, N. 2006. Türkiye'de Sınır Ticaretinin Gelişimi, Ekonomik Etkileri, Karşılaşılan Sorunlar ve Çözüm Önerileri. *Zonguldak Karaelmas Üniversitesi Sosyal Bilimler Dergisi*, Cilt:2, Sayı:3, Para Dergisi, "Kent/VAN", 2 Mart1998.

Polese, A. 2012. "Who Has the Right to Forbid and Who to Trade? Making Sense of Illegality on the Polish-Ukrainian Border." In *Subverting Borders: Doing Research on Smuggling and Small-Scale Trade*, edited by B . Bruns & J. Miggelbrink. Berlin: VS Verlag.

Roitman, J. 2006. "The Ethics of Illegality in the Chad Basin." In *Law and Disorder in the Postcolony*, edited by J. Comaroff & J. Comaroff. Chicago: University of Chicago Press.

Schendel, W., & Abraham, I. 2005. *Illicit Flows and Criminal Things*. Bloomington: Indiana University Press.

Şenoğuz, P. 2014. "Ahlaki Ekonominin Sınırları: Kilis'in Kayıtdışı Ekonomisi ve Yeni Zenginleri." *Toplum ve Bilim* 131, 105–135.

Wilson, T. M. 2014. "Sınır'ı Tartışmak: Yuvarlak Masa Söyleşisi." *Toplum ve Bilim* 131, 11–32.

Yükseker, D. 2007. "Shuttling Goods, Weaving Consumer Tastes: Informal Trade between Turkey and Russia." *International Journal of Urban and Regional Research* 31 (1), 60–72.

3 Ethnicized borders: Old boundaries, new situations

Border experiences of the ethnic groups

As a small border town, Hopa has always had a lively economic life with the harbour, which has been very busy for years. New opportunities occurred by the reopening of the border at the end of the 1980s. Attraction to liveliness in economic life made Hopa a town where different social and cultural groups lived together. Some of these groups weren't unsettled and had temporary trade relations. The other part of these groups settled in Hopa, but to survive in the town center was hard for them. Population in Hopa is about 40,000 including Lazis, Hemshins, Georgians, Loms, Kurds, and Turks. However, Lazis and Hemshins are the main part of the population. Today they are the most significant groups in the economic, social, cultural, and political life of the district.

The dynamics that the border economy creates reflects itself mainly within intergroup relations of ethnic identities. The forms in which the ethnic groups participate in the economic life that appears in the border regions and the way in which they experience that life, the new dynamics that these relations creates within the groups, and the relations between the groups presents us the points at which the economy and ethnicity intersect on the border. This corresponds to a mutual dialectic and dynamic relationship. In the same way these intra- and inter-group relations that we are talking about are fed by this economy, how and in what scale these economic activities continue is also fed by these forms of relationships.

Observing the relations between Lazis and Hemshins, it would not be inaccurate to take the reopening of the border gate as a milestone. Lazis who settled on the coastal area in the period before the reopening of the border owned both the land and the harbor as well as all positions in official institutions, and thus they were the only dominant group in this coastal area. Hemshins, on the other hand, lived in mountain villages away from the coast, were engaged in animal husbandry, did not go down to the center, and did not enter into relations with official institutions. The dual structure that emerged in economic life after the opening of the border was based on factors such as land ownership, exclusionary/oppressive forms of relations as noted above, and the fact that economic advantages emerging from the border were more for one group. As Barth (1969) states,

Changes in established relations between social groups often catalyze identity formation. That is, when one group considers its interests to be threatened by another it moves to strengthen the divisions between the groups by reinforcing the perception of differences and by constructing boundaries to identify common interests with other groups.

This is what was experienced in the Sarp border region between the Lazi and the Hemshin. Hence, after the opening of the border, the Lazis used the advantage of owning more land and became owners of real estate, opened starred hotels, or at least continued as tradespeople and got involved with this economy. Hemshins, on the other hand, sold the trucks by which they conducted their transport businesses and bought semitrailer trucks on credit—today the number of semitrailer trucks in Hopa is over 800—and improved their businesses. Using the language advantage, they became the executers of the trade with Armenia over Georgia, although that is officially invisible today. On this side of the border, in contrast to Lazis who carry on a trade that is less risky, they became the basic component of transborder trade and had to face all the risks that come along with this trade. But at the same time, Hemshins benefited the most from the economic advantages of this trade. This situation began to threaten the centuries-long domination of Lazis at the center, so that othering discourse between these two groups began to take more space in everyday life. Martinez (1998: xviii) says that all border inhabitants have a shared "border experience." According to Flynn, border people do not try to deal with shared experiences with these challenges and processes in the same way everywhere (Flynn 1997, 312). That is, these processes are experienced and perceived differently in different border areas. Our study has shown that experiences and perceptions of different ethnic groups differ. Instead of a communitas to commonize all border people in the liminal space, a life emerged in Hopa in which each group created its own "communita." This allows us to observe the differences in ways of involvement with this economy, the differences in the effects of the border economy on the relations within these groups, and the new life dynamics emerging at the intersections of border economy and ethnicity.

Boundaries from the past

Lazi and Hemshin who have been living together for almost 300 years

> have a number of features in common. After a long history of Christianity, both were converted to Islam during Ottoman times. In modern Turkey, both constitute minorities, albeit unrecognized. They are well integrated into the modern Turkish state and identify readily with Islam. They share many cultural traits with each other, as well as with other groups with which they interact... Members of both groups claim complete ignorance of their Christian past, although indications of such knowledge about the

other group are embedded in occasional mocking references such as the Hemshin calling the Lazi converted Mingrelians (*dönmüş Megrel*), to which the Lazi may retort that the Hemshin are converted Armenians (*dönmüş Ermeni*).

(Hann 2007, 339–40)

Even though both groups are trying to marginalize and in a sense humiliate each other by emphasizing their Christian pasts, the truth in today's conditions is that both of these groups are Islamicized and have constructed their lives accordingly. The thing that determines their existence together today is the form of their participation in the economic life and the dynamics that this process creates. In the words of an Hemshin interviewee,

Hemshins have descended to the seashore during the past 20–30 years. Hemshins are people who live in high altitudes, semi-nomad people that we can call mountain dwellers. Lazis are people who are closer to the sea culture, people who have lived in the shores more. Now this is the concrete situation. Hemshins have left the places that they lived in the last 20–30 years and came to enter the living spaces of Lazis. The Lazis whose economic conditions were good have already left these places, sold their lands, and went to big places like Istanbul. There is something in that sense between other Lazis who stayed and Hemshins. Lazis don't want to lose the rule they have over where they have it. Hemshins are in a condition that they have to live here. There is no life in the villages anymore, No one is cultivating their lands. They go for three3 days to collect their tea and come back. The lifestyle before is abandoned. No one goes to high meadows, there are no more doing high meadow activity. The economic situation has differentiated completely, and the lifestyle has differentiated. We are going to live here from now on. It is both the Hemshins and Lazis who have to live here too. But Lazis were living here before. Hemshins, on the other hand, like Batumis, would do their shopping and then return to their homes again. Maybe friendships at that time were stronger and more naïve now. However, we won't stop coming and coming.

(MH_IW-22)

The two groups are still clearly demarcated. In districts that have both the Hemshins and the Lazi population, it is common knowledge which villages are Hemshins and which are Lazi (Hann 2007, 341). While they started to live together in the Hopa center, after the opening up of the border, they preserved and carried on their boundaries, this time by separating their quarters, tea houses, and restaurants.

In the past, it is known that there were close friendships between the Lazis and Hemshins who were living in close villages. However, there are also incidents that the old people tell about Lazi landlords treating Hemshins badly and enforcing tributes on them. Despite sharing the same lands and being neighbors for hundreds of years, the cultural interaction between them is still at a minimum level today. Despite being cultural next-door neighbors for years, persons who can speak both languages are almost nonexistent.

It is possible to talk about a duality that strikes the eye in every dimension of life in Hopa. This is to say that the public locations, commercial life, and political spaces are divided between Lazis and Hemshins, even though sometimes these boundaries are surpassed and they trespass into each other's space. The most fundamental discourse of these two groups is that "Lazis are pretentious and arrogant" "whereas Hemshins are coarse, from the mountains, swearing." In our interviews, the subject inevitably came to the Hemshini-Lazi separation and the effects that this has on their daily lives. Hemshins emphasize that Lazis are "city dwellers," "full of themselves," and "arrogant," and they say that they treat them as second-class citizens. Lazis, on the other hand, emphasize that Hemshins are rude, descended from the mountains, and talk about them as swearing while they speak. The othering expressed in words like this shows itself in the forms of not going to the same places, not doing any work together, or not belonging in the same political formation in their daily lives.

From the past until today: Spatial differentiations

We see the separation in the social life initially in the space. The spatial differentiation is a consequence of the forms of relationships that these two groups have developed throughout their history. The center has been held at the hands of the Lazi who have existed there for a long while, maintained the economic and political power at their hands, and been land owners. Central Hopa, the Middle Hopa district, is in Lazi hands. This region is at the same time the center of all the state institutions and shopping places, which means it is the "center" of the city in every sense.

Hopa is not a district that is suitable for habitation because of the geographical conditions, similar to how it is near the Black Sea in general. It has turned its face to the sea and its back to the mountains on which settlement is not possible. For this reason, the only suitable place for the Hemshins who settled in Hopa later was a dry river bed. The first Hemshins who came and settled at the center settled in a river bed that the Lazis wouldn't care to even look at. The later arrivals did not break this tradition either and settled in the Sundura district. Today, the Sundura district is like a symbol of Hemshin economic and political power. It seems that the Hemshins are also present at the Hopa center with its multistory buildings, shops, and paved roads.

> The shore regions belong to Lazis, and the hills belong to Hemshins. In Hopa, Ortahopa belongs to Lazis, and Sundura belongs to Hemshins. Ortahopa like we say among ourselves—the high-society segment is at Ortahopa, but the *varoş* are at Sundura… Lazi-Hemshini separation also exists among the people who live here. The thing about Lazi-Hemshini also exists on the side of the administrative folk. Now because that place

is generally owned by Hemshins, whatever that is going to be done is for that side. That is, for Ortahopa.

(FH_IW-3)

The spatial differentiation reflects the inequalities that exist between these two identities too. The Sundura district, which one Hemshini described as a shantytown (*varos*) and in which the Hemshinis are settled, is a district that was built upon a swamp and which still has streets that are not paved even today. Middle Hopa, on the other hand, in which there are shops and state offices and in which the Lazi are settled, has been the symbol of political and economic power. Even though Hopa has a small settlement space, it has been possible to observe the ethnic and class differentiation over space.

All of Hopa used to belong to Lazis in the past. Like three to five who are shopkeepers. For example, my father-in-law is a shopkeeper. He has a place at Hopa's seashore, for example, the place that we have our hotel. And he earned his money and then bought it. That means just like that, he is here because he is a shopkeeper, but both Hopa and Sundura and Ortahopa, all these places used to belong to Lazis. It is after a while when Hemshins discovered the city, that is to say when they started to return to the city, slowly they generally settled down in Sundura... Now you see the Hemshins who have been in Sundura until six or seven years ago have started to buy houses in Ortahopa. This time, there are those who left Ortahopa, deserted Hopa, and who went away among the Lazis.

(ML_IW-13)

The Lazis, who settled down at the seashore even before the drawing of the border, had to share these places with Hemshins after the reopening of the border in 1988. Spatial separation is still very visible between the districts inhabited by Lazis and Hemshins. Lazis live in Ortahopa and Hemshins live in the Sundura district. It is nonetheless possible, albeit undesirable, for these two groups to blend in certain areas of the town. Both groups prefer living their lives with the minimum communication possible:

My building neighbors are Hemshins, for example. I don't make a problem out of it. We haven't experienced any problems either, but it does become evident when you enter that society. For example, when you enter the society of my Hemshin friends, you realize the difference at that instance.

(FL_IW-20)

It is possible to understand the interrelationships that ethnic groups have with each other as a field in which the border trade has disturbed the balances. The balances have changed in these relations by looking at the locations of settlements. Hemshins, who gained power in the economic and political life of Hopa with the earnings that they obtained from border trade, are trying to move into Middle Hopa, which has become a symbol of this power and wealth. Lazis, whose economic and political power has weakened, are trying to keep their own existence by either migrating out of Hopa or othering the Hemshins with whom they are forced to live, together with discourse appropriate to the context.

> Now if I want to rent a house at Ortahopa… If I want to rent the house of a Lazi person, they ask beforehand, you see. For example, I went there, and my ethnic identity is not known at all. You are the landlord, I came, and I want to rent your house. The first thing that they ask is not what job you work or the like but are you Hemshin or Lazi. "We don't rent to a Hemshin." This exists a lot here in Hopa.
>
> (FH_IW-3)

There hasn't been a relationship based on violence in Hopa's social, economic, cultural, and political life between these two groups who have lived in places close to each other in Hopa but generally without having any communication throughout their history. A life the rules of which are determined—who will live where and how much they will be able to enter their own spaces—in which both sides follow the rules has gone on in general. As a result, a situation of conflict has not surfaced. However, the process after the opening up of the border has disturbed these balances and the rules that were followed even though they were not written.

> The place that they hang out most, the play house or the tea house as we call it, you go there to play gamble. You go to a Lazi tea house and you cannot find a single Hemshin. However, a lot of them 90 percent know each other. They are much closer friends outside of Hopa, but when they come here, they go to their own tea house, and the other one goes to their own tea house.
>
> (MH_IW-15)

Despite the fact that the relationships they have with each other seem to have increased after the opening up of the border, they have always continued to be the "other" for each other. They have continued to live in different districts and to go to different tea houses, restaurants, and they do not go into each other's houses.

> As far as I know, I don't know one single Lazi who has shot a Hemshin. I don't know of the existence of any blood feuds. I have never heard of it at all until today, but in social life, tea houses are separate, restaurants are separate.
>
> (MH_IW-2)

"Social contact" with the "other" (Yükseker 1993, 10) is an important determinant in the construction of the ethnic identity and continuation of the group identity. Ethnic groups define themselves by looking at the other. Lazis and Hemshins have also looked at each other for centuries, although their social contact with each other has been very low. These groups, who have been forced to establish more contact among themselves (translator's note) with the border even though they did not want to, have this time attempted to preserve their group identities by separating their social locations and developing new othering discourses about the other group.

Ethnic groups in work life

These two groups who live in the same geographical region did not have an egalitarian relationship because where they lived determined what jobs they did and what jobs they did their social lives, relationships and identities. Lazis lived in the sea shore and did fishing, Hemshins lived in the mountain villages and high altitudes and did husbandry. Lazis who owned land at the central districts and who were closely integrated with the state, which means with the government agencies, have established an oppressive/marginalizing and domineering relationship over the Hemshins who lived in the villages and did not get into contact with the government agencies unless they had to.

> [Hemshins] who own a herd slash a lamb, plant their corn or buy their corn and wheat while they are passing through a village, and say, "Aha, here are two lambs for you/" They sustain their livelihood and go on, and they have no relationship with the state either, neither on the basis of taxes nor by the laws. They have given one identity and said, "Take. We are going, and by parting from Hopa, we are climbing up the Georgian mountain. We come down from the Georgian mountain and climb up to the Bilbilan."
>
> (MH_IW-2)

For Hemshins, a situation of oppression had existed for many years. Interestingly, this oppression came from another ethnic group. They shared the same destiny of being subject to the hegemonic culture of the Turkish nation state. In terms of local dynamics, Lazis who held the political and economic power have continued this situation for many years as such.

The reopening of the border disturbed the longstanding status quo that had been in effect since the border was set and since Hemshins were deprived of their access to the fertile meadows of Batumi.

Hemshins started to migrate toward the center as the herding business started to diminish, and they were forced to find new types of jobs in order to survive. Transportation and freight were the end result of this process. Even though the Lazis tried to expel the Hemshins with rocks and sticks and by seizing the goods that they have brought, they couldn't overcome the Hemshins whose economic power had increased. This process had operated more slowly and more calmly while the border was closed, but it gained new dimensions as the new economic conditions that appeared after the border's opening created more opportunities for the Hemshins. Hemshins whose economic power increased started to make themselves present in the political arena too.

New dynamics that reveal with the reopening of the border

The Lazi who live at the center of the district benefited more from modernizing processes like education and advanced in jobs like civil service and trade. The Hemshins who were living in the villages could not take their share of these a lot.

> There is discrimination and the like in general in here. They have sent all the Hemshin girls to scholar places, and we have kept on going like sheep without realizing it, you see. We have entered a new job. They have settled here (in the factory) as office workers in the desk jobs. We came and, of course, were left out. They have given the cleaning job at the production. It is generally they who do the jobs like office work.
>
> (FH_IW-14)

Expectations and demands from the urban life of the Hemshins, who became effective figures in the economic and social life with the opening up of the border, have increased trade and every institution of the administration. However, Lazis who held these positions at their hands tried to continue this power that they had by favoring their own relatives and people they knew.

> Hemshins are very few in state bureaus in civil servant positions... for example, if there is a new hiring, if there are going to be ten people hired, eight of them will be Lazi, because the higher ups are Lazi, and they hire those who are of their own.
>
> (ML_IW-10)

For these reasons—both the low level of education possibilities and Lazis being employed in state institutions—Hemshins have inclined toward occupations like transportation and driving.

> We have left herding in 1976. We haven't leave all of it slowly. Since 1976, we were carrying loads to Iran from the dock of Hopa. We were going to Tehran. My uncle, for example, left herding in 1978 and bought a lorry, and we have started transporting like that. That comes from the family, and we are still going on.
>
> (MH_IW-2)

The reopening of the border increased the opportunities for Hemshins to continue the businesses that they already had and to be able to earn more money out of it.

> The totality of the transportation business belongs to Hemshins, that is, 99 per cent of it. It is occupational, that is to say, it has become something that passes from father to son. We started this business working with mules. It is the Hemshins who carried the goods in Hopa to Artvin and to Ardahan with mules. You can go until the 1930s or earlier times, until the Ottomans. With mules we carried loads, carried goods, carried this and that, all with mules. We left the donkeys and bought mules. We left the horses and came. This is just how we are going.
>
> (MH_IW-2)

Another situation that provided an advantage for the Hemshins after the opening up of the border is their language. They have especially used the advantages of speaking the same language in the trade that is being done with Armenia, and they earned bigger profits from both transportation and transborder trade.

> It has provided an advantage to us as Hemshins. We have made commerce with people who are in any case talking the same language with us. We have gone there with nonexistent capital, and we were able to bring back goods just because we knew the Hemshini language. There was an advantage like that. We never experienced a disadvantage because of our being Hemshins.
>
> (MH_IW-22)

A huge majority of Hemshins deny the fact that they are Islamized Armenians because of several reasons. The first is that they were Islamized a long time ago with the Islamization policy that the Ottoman Empire carried out, and a part of them forgot about their roots. This situation is more accurate for the BasHemshins who live in Çamlıhemşin and Rize-Pazar (Athena), whom I mentioned in the section "The Hemshins." A more important reason is the issue of Armenian genocide and the political problems with Armenia. The official ideology of the Turkish Republic which can be expressed as "one language, one nation," caused Hemshins to deny this origin of theirs in order to avoid being subject to cruelty. The reopening of the border and the non-official trade that is done with Armenia over Georgia turned into an advantage for Hemshins.

> Of course, it is not only Georgia. There are also other borders beyond that. Hemshins do a lot more business in the Armenian region originating from being Hemshin and related with the language. There are also those who say they are Armenians. After all, there is also an advantage that it

brings, and they use that. They go and are more comfortable there. I say this because they are comfortable Hemshins and do more trade there.

(ML_IW-10)

Lazis, on the other hand, being settled in the seashore for centuries, benefited from holding the land ownership. Only one of the big starred hotels belongs to Hemshins, and all of the others belong to Lazis.

Now the Hemshin people know commerce, know transportation very well. Now everybody has a quality. The business of those people has been working cars since the old times. They used to work toward Iran beforehand too. They carried loads with trucks and the like. They had the first place in transporting. The job for the Lazis is generally fishing. The commerce of Lazis is a little bit of everything, not transportation but shop keeping, operating markets, and things like that.

(ML_IW-13)

Despite our interviewee's claims that there is no discrimination in business, official business partnerships of the two ethnic groups exist in only two or three companies. They generally prefer entering into an official partnership with their siblings or friends. They do business together in only the short term and for temporary jobs.

I have done partnerships with Lazi friends who are not Hemshins. We have done partnerships with lots of goods, we have done travels together, and there have been friends with whom we have gone for a long while... Of course, there is shutting each other out. We cannot say that there is not. If we say we now have relationships with 100 people, there would be at most five Lazi among them. And this is a good ratio too.

(MH_IW-22)

Economic and political dominance of Lazis, which has lasted many years, seems to have come to a halt by the balancing of power and thanks to the border economy. Oppressor/oppressed relationship between the two ethnic groups gained new dimensions with the equalization of economic and political powers. However, the duality, if not segregation, in the lives of the townspeople seems to have continued despite the balance in their competition for economic and political power. In other words, balanced power relations between these two groups have not provided a fully integrated community life.

These new forms of living and forms of establishing relationships that the border brought engendered new exclusionary discourse in both groups toward each other. The othering has continued by emphasizing the occupation or the family life of the other party. The example of Teksas Avenue illustrates the new divisions that appear in the ways in which members of each group enter into the new economic life of the town.

Teksas Avenue: The "outlet" of border economy

Teksas is the microsized case for ethnicized life and space in Hopa, which we discuss in this section. It will be appropriate to describe Teksas Avenue as an "outlet." An outlet, at least in its widespread usage in Turkish, is the name given to the stores that sell goods (usually clothing) that cannot be sold at regular prices, export leftovers, clothing from the previous season, or goods with defects. The border region exhibits a similar quality of being an outlet for the border economy. The transit role of the town makes it an outlet for large-scale commerce and smuggling. Profit is lower for the middlemen and commission merchants. Consumption, on the other hand, is of the leftovers of the large-scale circulation of merchandise.

Dumlupınar Street acquired the nickname "Teksas Avenue" after the opening up of the border gate. Bars and hotels on this street are venues for prostitution of women from various former Soviet Union countries. Clients who make deals at the bars continue the night in the hotels. These bars are in the basement floors below the lobbies. Bargaining is made in the lobbies. Occasionally, disagreements in bargaining end up as fights involving guns and knives.

> There are seven to eight bars on this avenue... in the middle of the market. But there of course, we are not backwards minded. Let there be everything, but it goes a little bit absurd within the market place. This kind of business can be done at side spaces of Hopa, not in the center. I think, according to my idea, it would be better I mean. I don't let my wife pass on this street. For example, I also told my children not to use this avenue. Oh this place is just like the Teksas Street that you talk about. In that regard, that is, I tell my family and children to not pass there, because there are drunk people who come out from there. Let them be at a side space instead of inside this market place. Let them be 10, let them be 20, but let them be a little bit away from the market place.
>
> (ML_IW-8)

Those who run businesses on this street are mostly Hemshins. There are no Lazi shop owners to be found. Based on the claims of my interviewees, one reason for this might be that the mayor, himself a Hemshin, was backing other Hemshins by renting out these shops to them. Another reason, on the other hand, is that Lazis are the ones who were doing the first-class businesses of the border economy in Hopa. The big starred hotels belong to Lazis. Lazis are the ones who own the land at the center of Hopa and who settled in for a long time.

> Again, as a result of the gate, one of our avenues, the Dumlupınar Avenue, is more Georgian in the position of dumps, not starred hotels, but hotels in that position. They hang out in beer houses, and the women

cheaper and in more trashy conditions. There are worse, that is to say, the ones in the hotels do not seem to the eye very ugly, but those places on that avenue seem very ugly... Because in there, those kinds of people hang out a lot, but then again, we are speaking, if it is making a statement, let's say it is the shopkeepers in that street are too those in the front the people who live on the right and left those among who inhibit the passing nine out of ten are unfortunately Hemshins. Let's say it is nine of them. I would be an exaggeration to say 100 percent. There is a truth like this. The businesses there are theirs more or less. These are in the form of beer houses, but they have surpassed beer house thing. You see, in certain parts of them, in their sections like rooms, how should I say? From time to time, I have gone there for certain reasons. There are sections where they sit to market themselves. The women sit there although that place is a beer house. Those places are forbidden places, actually, but these places are places that the police force overlooks. There shouldn't be those kinds of people in that form in those places. There are people who go who are old without money or with little money. It is a dumpster exactly like those in the movies.

(ML_IW-10)

The fact that all the business places in Teksas Avenue are owned by Hemshinis has deepened the Lazis othering discourses that had historical roots and which were based on the occupation and lifestyle. The new sex industry created a new discursive degradation of Hemshins, based on the amoral quality of the jobs undertaken in this industry. These discourses are also connected to the lifestyles and family relations of the Hemshinis.

Changes in group relations

Within the modernization processes, together with urbanization, a transition from primary relationships to secondary relationships had been experienced in here too. While the Lazis experienced this process earlier, the process sped up a lot with the Hemshins following the opening up of the border.

It wasn't like that before, that is to say, I remember now from my mother, for example, there was corn planted in places where the tea did not come out very densely in the gardens. There were processes of collecting the corncobs at one place and separating them from their cobs. In that process from the villages, across even from Kemalpasa, there were people who came there in solidarity. There were common things like that. In there it was both girls and boys, one across the other, songs and ballads, etc. There were situations like these. Now there aren't because that production has ceased to exist. They have become tea gardens instead of corn.

(MH_IW-1)

One of the important dynamics that the opening up of the border created in the daily life is the disintegration of social solidarity that was created with cheap labor. Collecting tea, harvesting corn, or similar duties were done through collective work with the solidarity of neighbors and relatives, but a majority of these jobs are now done by Georgians. Georgian men start waiting in the center, which is called "Georgian Market," from the early hours of the morning, and those who need workers come and make agreements in this market.

> We used to have our old collective labor system with collaboration. That finished completely. Now when Harun is sick and there is tea, he is sick or is going to build a house, build a road, go and help him out? It is not possible. Harun will go and hire the Georgian citizen as worker and deal with that job. The collaborative labor system was producing collectively. You see, the corn fields were harvested all together. You see 13 people entered a field. A field was finished three days later, and another field was being gone to. He would go to him, help him out. That is to say, when building a house, they would never pay a worker's salary. They would only pay the salary for the journeyman carrying and load-carrying jobs. People would carry rocks on their backs and build houses.
>
> (MH_IW-2)

The opening up of the border changed the way in which people form economic and social relationships within the group and their intergroup relations. The respect and trust that they had for each other diminished within these new forms of relationships that the border created. "The economic relationship at the border is a commission economy. And this is the lowliest form of the capital with the least character. Everyone who has entered these relationships has become characterless" (MH_IW-39). The people, who would give their lives for each other under many conditions, were not able to keep the words that they gave each other under these economic relationships. As a result, the relationship of trust between them was disrupted. When the border was first opened, there was not an economic relation that was professionalized, that left the human relationships outside. As a result, in this unregulated and unstable environment, everybody ended up in a place where they seemed like they were scamming each other. The people warned each other by saying "Don't do partnerships or commerce with anyone" (MH_IW-39).

New conflicts within group relations

As we mentioned above, the forms in which Lazis and Hemshins participated in economic life exhibited differences. While Lazis owned hotels or big companies or at least did shop keeping (using the advantage of their greater ownership of lands), the Hemshins continued what they have been doing for a long time and did transportation: being a driver or opening up a transportation company and buying lorries. While Lazis were utilizing the resources that already existed, Hemshins participated in border trade by the material support that they received from their families and by collecting their resources together with their relatives.

> Look and see this building. I am saying this as a person who is 60 years old, these places where we are sitting were swamp. There was only (...) market. These buildings over here were made by that money. There are 300 businessmen in Hopa, but I can only see ten of them. There were not 5 to 10 lorries in 1974, but now there are over 1,000 lorries. I am asking you, sir, where did these 1,000 lorries come from? They were bought from some people's accumulations or with the help of friends and acquaintances.
>
> (ML_IW-6)

The effects of this process on the relations within the group came about in various ways. The uncertainties and risks experienced during the initial period, which I explained in detail in the chapter "The border economy," caused the money that was invested to be lost and the relationships with the relatives to be worn out.

The things that one of our respondents (whose father has disappeared on the other side of the border) said displays both the hardships and risks of the border trade and the trade that is always done with acquaintances and relatives. Not just in the Turkey-Georgia borderland, but almost all "cross-border economic and commercial activities are often based on preexisting networks of kinship, friendship, and entrepreneurial partnership that now span both sides of the border" (Baud and Schendel 1997, 230). The thing that this respondent of ours said at the same time is an example of how those from one's own group are preferred while trade is being done on the other side of the border and that partnerships are established with family or relatives.

> There was an Armenian who was living in Georgia. He first worked with us for a year or so. Then he said, "There is pure copper in Armenia." They established an office and stuff in Georgia. We gave that guy $100,000 or so or to be exact. My father gave it. He was bringing the commodity and selling here. He was working regularly, I mean. Later on, I wondered how it happened but we gave those people $100,000. Fifteen days later or so, he said like it was caught in the border while it was coming, like there was a missing document. This guy in a month or two said that the commodities were caught, and I couldn't do anything. That money went away like that. After a while, my father started a company again through his own struggle... Then he said, "See that I am earning money, let me do business with my relatives"... they lost that money too... I mean, we experienced a lot of hardship in the economic sense.
>
> (MH_IW-11)

What one of our interviewees (who was born on the other side of the border and settled in Hopa after the opening up of the border) experienced shows the disintegration of the relationships within the ethnic groups themselves especially because of the uncertainties that the economic conditions created.

> We had a relative here. I don't want to tell his name. It was him. You see, he had companies. We later found out that his company went bankrupt. At that time, of course, we did not know about it. He has a house in the Sundura district in Hopa. Five stories not completed. Its bricks are laid. It turns out that it was under mortgage. We did not know about this either. The night that we came, that relative of my father-in-law said "Uncle, you give me the money. On you I have houses in Hopa for at least six months. I will give you two apartments. I will stand as I have given you your money back. You will have a house for yourself and one for your child."
>
> (FH_IW-16)

The interviewee from whom we quoted above in the end neither had house nor could take his money back. The relatives that he talked about went bankrupt and could not pay his money back. Today, he says, he and his relatives avoid each other because of that incident. This process cannot be named as a conflict but rather a disintegration of the relations within the group. Certainly, this process is also a result of migration from village to the city. However, the opening up of the border brought a rapid integration to the economic life and accelerated this process.

> I don't remember now. I haven't gone to any of my relatives' houses in the last one or two years. It is not like my paternal aunt is going to come to me. It is me who has to go. We used to go in the feast times, and we would travel around, see. There was a coming together, a cohesiveness. Now people even in the streets, we cannot say hello to each other sometimes. It is as if when you say hello, somebody will ask for something. Because there are economic hardships, something happens to everyone.
>
> (MH_IW-11)

The process operated differently for Lazis. Lazis who were working as civil servants in governmental offices or had been shop keepers for years in general stayed away from jobs like transportation and driving, and as a result, they were less affected by the risks that the border economy created. During the period when the entertainment sector boomed, a number of Lazi families migrated from Hopa. That's not to say that Lazis were immune from the drastic transformations brought by the opening up of the border. Rather Lazis, as the dominant economic group of the town, in a way experienced a similar disintegration caused by urbanization and modernization at an earlier stage than the Hemshins. Thus the Lazis whose extended family ties had long been dissolved experienced the impact at the level of the nuclear family. The money the Hemshins lost in business ventures involved their extended families, siblings, and relatives. The relation of trust was worn out with commercial failures. This situation was aggravated by new problems in family life: the role of the entertainment sector in spoiling marriages and increasing cases of infidelity, divorce, or double marriages. This situation will be detailed in the chapter that follows.

New otherization: Intergroup conflicts

The Turkish-Georgian border can be considered politically stable primarily because of the stable territorial boundaries of the Turkish Republic accompanied by the construction of the political identity and unity. There is no problem at the level of integration of these two ethnic groups into the state. Both groups perceive themselves as parts of this political unity. Despite the lack of identity claims at a larger political level, there is a hierarchical structure that has survived from the past between these two groups that shape their relationships with each other.

Two groups perceive each other based on the historical factors that created differences. Lazis see Hemshins as "coarse, uneducated," and Hemshins perceive Lazis as "snobs, pompous." The words of our interviewees below are a clear expression of the discourse that enables these boundaries to continue.

> I mean, the Lazis regard themselves as a little bit on a higher ground on this matter. They say, "We are more cultured." I mean, I know it. Hemshins used to be a little bit more something in the past, foulmouthed and the like. That's why they didn't like them much. They are coarse people, herding people, those I call Hemshins. The majority of them were herders in the past, you see. They used to stay in the mountains. A man can be coarse when they descend from the village, you know. Their manner of speech and other stuff, they used to regard them as coarse. It doesn't work without the Hemshins either. They also know themselves too that it doesn't work.
>
> (FH_IW-7)

As can be seen in the quotation above and the emphasis of the Lazi respondent given below, exclusion has become a case of "being cultured" and "not being cultured." Here, "being cultured" means being from the city and having an education. In the subtext of the emphasis of Lazis, the anger from the economic and political power that has been lost and the desire to continue the Lazi identity by underlining the differences can be read.

> The culture of Hemshin is a little bit broken compared to the Lazi. I mean weak... all of the cultures of the Hemshin—economic, familial,

social—are all weak... Take ten Lazi and speak with them, and there won't be three mistakes, but take the Hemshins, and nine of them or eight of them will make mistakes. They will use 20–25 curse words, even when having this conversation here, without regard to your being a lady.

(ML_IW-6)

As we have mentioned above, these boundaries, especially visiting each other's houses, is low or nonexistent. Even friends from the two groups do not visit each other's houses, although they mention that they meet outside and "hang out." The reason for this is that the house, as in many other cultures of that geography, is seen as the most private place and the sanctuary of family life. These boundaries have become unbreakable and unassailable and are internalized to such an extent that there has been almost no change in the thoughts of the Lazis and Hemshins who live in the same apartment building next to each other.

I am saying you can have two cars and 100 lambs. You have to sell them if you are going to live in a building. Learn how you are supposed to live there before living there. You say that, but they enter from here and exist from there. Laundry is done on the third floor and hung up, but there is laundry that is hung beforehand. You are going to warn them as a woman. You should say, "My friend, I am going to hang my laundry up if your laundry is dry."

(ML_IW-24)

Lazis who had to live in the same place with the Hemshinis whom they have seen as villagers, uneducated, and rude deepened this discourse they carried over from the past by emphasizing the weakness of the family relations of Hemshins. The interviewee saying "they won" (whose words we quote below) is exemplary of both the anger of having lost the places that once belonged to them and envy of the improvement in the economic power of the Hemshins that they have seen as below themselves until this day.

[Giving the name of one Hemshin man] He is going to say the things that I talk about with curses... Why? It comes from the family. When I came here during the 1950s, Hemshins had a bakery. Lazis wouldn't do it, because it was hard work. There is a district called Kuledibi. They would come with donkeys from there to here. There were no cars, you see... They would sell wood. The only things they bought were sugar and salt. All was in the garden. There was nothing else. Now the owner of the house that I live in is a Hemshin. He is a restaurant owner. He is a man worth a trillion. They won.

(ML_IW-6)

A Lazi woman's daughter is married to a Hemshini and says that her daughter is experiencing problems with her husband because of the "Russian events." She says, "What can we say? We can't say anything. My daughter is 29 years old and has a son. If she is leaving, there is the kid. She can't leave the kid. The husband is not worth it either." When she is asked, "Doesn't her mother-in-law say anything?" she emphasizes that in fact the family is corrupted by saying "No, they are the same. All of them are same" and adds "The bride of the brother-in-law of my this daughter is also Russian. He has three children" (FL_IW-5).

> Sarp gate had a huge contribution to Hopa in economic sense. It disturbed the family ties. It is correct, but which one was disrupted? It hit the ones whose family structure was already disrupted since yesterday. They found money a little bit. Sundura district is the biggest district of Hopa. A guy in here had a secondary marriage, the one who is giving and the one who is taking. Are all okay with it. Those whose family culture is weak, those who do not look at their families in a healthy regard, I can never project the mistakes that people do in Hopa.
>
> (ML_IW-6)

The necessity of these two groups having to live together has continued their existence by having the minimum relations with each other while living in places close to each other for many years. This has brought about new boundaries. In order to continue their group identity, Lazis have created new labels for the family lives of Hemshins. According to Lazis, Hemshins whose family lives were already spoiled have entered into "amoral" relations that the entertainment sector created, double marriages and divorce. Infidelity became common among the Hemshins.

> Hemshins are a little bit relaxed on this subject. They have more of the secondary marriages with the Georgians, but the social impact, the psychological impact, is not as much as the Lazis, and they are a little bit relaxed, more at ease... For example, those who have had their psychologies disrupted, who have left, have always unfortunately been the Lazis I am saying because of this kind of relationship. They are a little bit relaxed. They have thick ribs.[1]
>
> (ML_IW-10)

The words that one of our Lazi woman interviewees used for her gender-mates is an example of ethnic identity surpassing gender identity, and it includes harsh scorn about womanhood.

> Hemshini women do not interfere a lot with their men. They say, "It is okay as long as he brings me money," but there is no such thing. As Lazis they never accept. Lazi women say "there can be no one other than me," but I see the mother of my friend. Her husband's whereabouts are not known, but she says it is good as long as he brings money.
>
> (FL_IW-21)

Hemshins are trying to develop a counter discourse to all these new forms of othering discourse that come from Lazis by claiming that they are warmer blooded in nature and that they give more importance to family relations.

> No Hemshins express themselves as individuals within the family. Expressing is thought of as a little bit of weird. Socializing within the family is more on the forefront. Even if it is your spouse, going out together is not a lot. With your mates and friends all together, even if they are going to stroll, they are going to stroll together. The nuclear family stays together all the time with kids and such. For the mothers and fathers here, those people's individual behavior is like not caring for the family a little bit.
>
> (MH_IW-15)

Despite Hemshins gaining more economic power and their presence in the political and social life of the town can be felt, the hierarchical structure that exists between them and Lazis for hundreds of years has not completely broken down. Economic and political empowerment of Hemshins, contrary to expectations, has created new discourse, disdain, and scorn.

These two groups, which had to reside together in many locations as a result of the new dynamics that the opening up of the border created, developed new othering discourses that are in line with these new dynamics in order to continue their group identities. Lazis have developed these discourses over the "defunct" family lives of Hemshins, whereas Hemshins developed them over Lazis being "haughty" and "cocky."

Marriage patterns of ethnic groups

According to the primordialists, the fundamental basis for ethnic identities is the blood relation, and the survival of this identity is connected to the survival of blood relations. For this reason, group identity is attempted to be continued and preserved by always making marriages within their own groups. This situation is also valid between Lazis and Hemshins. These two groups have severely refrained from exchanging brides, which is to mean establishing blood relationships in order to continue their existence in the face of the other group which they live together in the same lands in order to keep the boundaries in place.

Today the number of marriages between the two groups is still small in number. And they continue to legitimize this with the cultural differences with each other. Hemshins claim that they do not discriminate against Lazis and say that Lazis are the ones who will not give their daughters to them. Lazis consent to this situation too. According to one of our Hemshin respondents,

> [Lazis] see Hemshins as backward. Uncultured. They regard them as if they are in a level below themselves. They see them as people who have

worked in the villages, who worked in houses, who are oppressed. A Lazi daughter is a lady. She doesn't work in the fields or gardens. Working in the fields and gardens is not something bad, after all. They regard Hemshin daughters like that. Their own daughters will be civil service workers, will be comfortable, will be a lady, won't be oppressed, just like that I mean.

(FH_IW-14)

Hann (2007, 343) states in his article titled "Hemshinli-Lazi Relations in Northeast Turkey," "Most informants, both Hemshin and Lazi, insist that intermarriage took a one-sided form; Hemshin brides were taken by Lazi men, but no Lazi girls married up to Hemshin villages." This is a situation that is also voiced by our interviewees.

Lazis would not give their daughters to us Hemshins with any ease. Of course, there are those who are married around us, but most of them are runaways… and those are like when a Hemshin man's occupational situation is really good. His income, etc. Like money, you know how money is a little bit at the heart of this deal, but other than that, there is no such problem with the Hemshins.

(FH_IW-3)

A Lazi bride is not a problem for Hemshins, however giving a Hemshin daughter to a Lazi is very hard. A Hemshin taking a girl from the Lazi is another kind of difficulty. I mean, Lazis are a little bit more despotic on this subject. I am saying this as a Lazi, you see.

(FH_IW-4)

The most fundamental reason for not taking or giving girls and for within-group marriages to continue is the pressure of the group on this subject.

Of course, for example, in here Lazis don't give their girls to Hemshins, but a Hemshin's daughter comes with two arms opened and comes as if Lazis are superior. But Lazis certainly don't give girls, but girls come from there… But now, for example, if I give the girl, they will say, "Why did you gave the girl?" just because of that, meaning out of pride.

(FL_IW-20)

Similar to those in the interviews that Hann (2007, 343) made, Hemshin girls being stronger and more studious is also given as a reason in our interviews for Lazis taking girls from Hemshins but not giving away their daughters. "Their daughters used to be better nourished, and their healthy lifestyle in the high pastures made them sturdy and strong and therefore desirable as wives for the Lazi. Both groups stress that Hemshin women have always been hard working."

The majority of the marriages could only be possible not by the consent of the families but rather through the insistence of young couples. When we asked the question, "Would you allow it if your daughter wanted to marry a Hemshin person?" one of our Lazi interviewees underlined that a Lazi girl in her right mind would not want to marry a Hemshin anyway by saying,

> I wouldn't give her, but I wouldn't interfere either, but I don't think she would marry. She has passed that age. I think she would not love. But I am again strongly emphasizing that every Hemshin is not the same. There were no marriages at all before. Now the girl can go herself too. If she loves, we can give her, but we don't give daughters through arranged marriage. If my daughter loves a lot, then out of necessity we would give her.
>
> (FL_IW-21)

A Hemshin interviewee emphasized that they are regarded as "inferior" by the Lazis, and for that reason, they don't want to give their daughters. She added that it is not possible to create marriages through "normal" means.

> Lazis don't want for us to visit each other. Hemshins don't have such a thing. Hemshins don't make any discrimination. Lazis don't want it much. You know how they say "second-class person" all the time? They see themselves as superior. Even in woman-to-man relationships, they don't take a Hemshin girl as a bride, or they don't give their girls to Hemshins. It happens once in a thousand, and that is either if the girl is a runaway, or if they insist a lot, meaning "I will take him in any case" if it is to that degree. Otherwise, if she goes to a normal family, okay, then she loves him. Let's ask for her, and they'll give her. There is no such thing.
>
> (FH_IW-12)

Even though some of the boundaries between these two groups have eroded after the opening up of the border, behaviors and attitudes toward intergroup marriage have not changed at all.

Premodern patterns of marriage yield intragroup and sometimes village endogamy, an extreme form of which was close-kin marriage. In practice, out-marriage has probably always been an option (Hann 2007, 345). Today, although Lazi-Hemshin relationships increased in many areas of the town's life, neither group affirms Lazi-Hemshin marriage. Even though they establish certain relationships, marriage is not one of them. This shows that overcoming symbolic boundaries between ethnic groups is often more difficult than overcoming physical boundaries, and moreover, as in our case, physical borders drawn between states deepen existing symbolic boundaries between border people and even create new ones. Especially if the same ethnic identity groups live on either side of the border and establish mutual economic or social relations over the border, this situation reinforces the group identities of those belonging to the same group while it further deepens the boundaries with the other. Depending on the networks of relations an ethnic group establishes with the border during the use and distribution of new resources that emerge with the border, not individual consequences but consequences that affect the group as a whole emerge.

Note

1 "Thick ribbed" is used for Armenians and more in order to insult.

References

Barth, E. 1969. *Ethnic Groups and Boundaries: The Social Organization of Culture Difference.* Boston: Little Brown.
Baud, M., & Schendel, W. 1997. "Toward a Comparative History of Borderland." *Journal of World History* 8 (2), 211–242.
Flynn, D. K. 1997. "We Are the Border: Identity, Exchange, and the State along the Benin- Nigeria Border." *American Ethnologist* 24 (2), 311–330.
Hann, I. B. 2007. "Hemshinli-Lazi Relations in Northeast Turkey." In *The Hemshin: History, Society and Identity in the Highlands of Northeast Turkey*, edited by H. H. Simonian, Abingdon: Routledge.
Martinez, J. O. 1998. *Border People: Life and Society in the U.S.-Mexico Borderlands.* Tucson: University of Arizona Press.
Yükseker, D. 2007. "Shuttling Goods, Weaving Consumer tastes: Informal Trade between Turkey and Russia." *International Journal of Urban and Regional Research* 31 (1), 60–72.

4   Gendered borderlands

We see that studies on gender at the borders are few, and existing studies often focus on the subjects of the entertainment sector and prostitution. This is understandable as "border zones have been widely reported as providing opportunities for illicit sex" (Donnan and Wilson 1999, 91), and the situation is also quite applicable to the Turkey-Georgia border. However, if the issue of borders is about the bodies of women, various dimensions of the issue should be addressed and elaborated. One dimension consists of "the symbolization [of the borders] as a woman's body whose honour should be protected" (Delaney 2001), and another dimension consists of the commodification in these areas of the bodies of women who come from across the border. A point that is lacking in border studies focusing on the phenomenon of illegal sex and prostitution concerns the border experiences of local women. Yet as the border penetrates the social, cultural, political, and economic life in the areas where it exists, it touches the lives of the women settled in these areas the most.

The "border is honor" discourse, the bodies of immigrant women victimized for border economy, or the shattered lives of the local women cannot be considered as separate from each other. All these are manifestations of patriarchy reproduced by sexist discourse in border areas.

Addressing the process of "the symbolization of the entire country as a woman" and thus the presentation of the protection of the borders of the country as a "matter of honor" in the process of Iran becoming a nation-state, Najmabadi (2000) emphasizes the Islamic roots of the concept of honor. In the process of becoming a nation-state, as the nation transformed from a religious community into a national community, the concept was detached from its religious meaning (*namus-i islam*) and found a national meaning. According to Najmabadi,

> borders drawn by wars in the 19th century with Tsarist Russia, the Ottoman Empire, and the British colony of India were imagined as the body lines of a woman: the body of a woman to be loved and to be devoted to, to be owned and protected, to die and kill for.
>
> (2000, 119)

It is possible to observe a similar conception in the process of the foundation of the Republic of Turkey. In Delaney's (2001) words, the land of Turkey "is symbolically a woman. Given the social structuring in relation to femininity, it becomes imperative that these lands are protected against destructive impacts from the outside." The protection of the borders of the country, which is sacralized by these symbolic loads, is a matter of honor, because the violation of these borders means the violation of its honor—that is, its woman. The perception of honor, especially in the Middle East, "as a value under the protection of the men" (Bağlı and Özensel 2011, 38) charges the duty of protecting the honor (that is, the borders) of the state or the men. This conception that perceives the country as a woman and its borders as her body and charges the man with the duty of protecting the borders is the reproduction of patriarchy within the modern nation-state.

Gendering of the country's borders also leads to the building of the discourse of those who violate the borders by sexist codes. For example, Yeğenoğlu (2003; cited in Özgen 2006) states that "one of the most distinct patterns of colonization is the womanization of the colonized nation, that the colonizer operates this process over the woman he representationally creates." According to Özgen (2006), "what was consumed was not the body, but the nations and the national covered by the image of the body." In the years following the dissolution of the USSR, there was a sex traffic in which young women from the former Soviet countries chose Turkey to stay temporarily and sometimes permanently. Özgen argues that this process should be read as "a way in which transborder trade of sexuality articulates to global capitalism." Nationalist discourse was developed over the bodies of the women subjected to trade, and another dimension was reached in gendering of the borders over the exploitation of their bodies at the same time. In this section, I address the experiences of immigrant women, and we will be able to see clearly how the bodies of these women are commodified in this trade.

As I noted above, one of the least addressed topics in studies on gender at the borders is the point of how local women living in border settlements experience the dynamics occurring with the existence of the border and what kind of relations they establish with the border. In the period when I worked at the Sarp border area, the point that attracted my attention the most was how local women became invisible in the chaotic environment created by the border. For this reason, I first address the border experiences of the local women.

Scattered lives of local women

At first sight, everyday life in Hopa is not different from a similar-sized town in Turkey. The men work, the women do the housework, concern themselves with taking care of the children, prepare house gatherings with tea-gold, and life seems to continue with all its normalcy and patriarchy. A closer look reveals the unique situation of a border town as a quite vivid and dynamic life. The lorry trucks are lined up one after another along the shore; hotels, restaurants, and small casinos follow each other; foreigners can be spotted by their looks, especially women who are comfortably and well dressed; visitors from other parts of Turkey can be identified with their license plates. The real effect of the border on the lives of the local people, however, requires research beyond observation. The first problem that comes up is the shattered family life. For these families located in a town, men go out to their jobs but do not come back or come back with another woman. The women who do not know how to hold their families together are lost in the middle of all this rumble. Children grow up in a family environment where their father is mostly not at home.

> The events of prostitution have been a lot. A lot of families have naturally been affected by this. There were a lot of men who went. Those who have taken and put the money of the tea there, those who have sold, forgive me for saying this, the cattle in their barns have gone there. The woman, you see, collects the tea, struggles and works, but I mean the man has turned out to be more dominant in the affair of prostitution. There have been cases of divorce, of course. Those who married other Georgians, those who had kids, those who left their spouses. There are those who went to Batumi and had secondary wives there, I mean. Those who had kids there. Their spouses did not hear about that at first, of course. They did later on.
>
> (FH_IW-4)

This form of life, which started with the opening up of the border and has taken a more tragic form with the increase of the entertainment sector, has displayed itself more in women's relationships with their spouses, relatives, and kids.

Ayşenur Kolivar (2000) in her book *Who Is Fadime?* uses the stereotypical name Fadime in the title, the female counterpart of Temel in typical Black Sea region jokes. She says that the Black Sea woman is

> stereotyped as gun holding, man like, getting whatever she wants strong woman on the one hand, an oppressed woman who has to work all day in the place of her husband who kills his time at the tea house, with loads on her back, a kid on her hands on the other hand.

There is "some truth like there is in all stereotypes." The difficult geographical conditions of the Black Sea, the placement of the tea-cultivated lands at the foot of the mountain, and the definition of the job of collecting the tea as a job of women are among the fundamental qualities of the life of a traditional Black Sea woman. This situation is not any different for the rural women who live in Hopa. It is the women who collect the tea, carry it to the collection places, and bring the wood from the forest on their backs. Even for those who moved to the cities, besides the daily "womanhood responsibilities," the tea collection job has continued its existence as a woman's assignment. There is the effect of the Black Sea man's having to work away from the city in the experience of agricultural production based on the woman's labor being more condense when compared to other agricultural regions. Even though the conditions are equalized and men have returned to their homes, these jobs are left to the women. One of my interviewees has expressed this with these words:

> Woman's word is the rule in the production. The tea is supposed to be collected… it will be harvested today, planted today, moved away today. The man does not interfere with that at all. A woman's word is the rule in these fields. That is true, because it is going to be the woman who will do the carrying, but when you cross to the other side of the job, a man's word is the rule. The woman collects her tea, spreads her manure, does her clean up, does everything, and suffers for it, but the right to use the money of the tea is the man's.

> (FH_IW-3)

One male interviewee was born and raised in Kazakhstan and settled in Hopa; his father moved back to Turkey with the "longing for the country of origin." He talked about this normalizing and acceptance.

> In here, the women work, and the men eat. The woman carries the wood too, cuts the wood too, carries the tea in a 50-kilo sack of tea in her back too, and her husband takes the money and goes, plays, gambles, and drinks his alcohol, and his own kids and kin eat the kale… Now when we were at the village, I said, "Let me do a couple bullocks," and I bought them. Like that, three or four of them, I have harvested grass. I will carry grass to the cattle in the evening, but the green grass is heavy, you know that, right? I am going too, and my head cannot be seen. This one here (referring to his wife) has the sickle in a plastic bag, and she is coming by, waving it around. The women have seen it and said, "Ooooh, what are you doing?" She said, "What's happened?" She said, "Are you crazy?" She said, "Why you are carrying the load? You should give it to your wife." Is it me who is crazy, or is it you? I cannot understand it.

> (MH_IW-30)

These women who have a say in the life of the village and who play an important role in the family economy lose their right of participation in the family life to a great extend after moving to the city center. According to what Kandiyoti (2011, 7) relates in the International Woman Conference that was done in 1975, modernization further deepened the cliff between the genders as contrary to the expectations, and furthermore, women were deprived of some power and privileges that they had in the agrarian societies. The logic that lies behind this is that as the forms of production geared toward a market economy (the necessary skills and, by extension, money) are transferred to the men, the value of the women's contribution declines. A similar process can be observed in Hopa: after the opening up of the border, the border economy became a source of sustenance for the families. Families migrated to the city center, their economic power increased, and they became more involved in the modernization processes. What followed were deprivations in the lives of the women similar to what has been asserted. While the tea that she collected and the goods that she acquired from the land had value beforehand, these goods stopped being the source of economic income for the family, and as a result, the status of woman in the family changed.

The women who are rendered obsolete in economic relations come into a position within the household where they only do housework, look after the children, and wait for their husbands to bring home money. The family economy that relied on the cultivation of tea and other sorts of production in the village lost this quality with the opening up of the border and became connected to the border economy. Consequently, the labor of women in the cultivation of tea became free but not internalized within the border economy.

Quoting from a study that Özgen made in Igdir (a small border city in the Turkey-Armenia border region) on sex workers (2006, 131), a non-governmental organization (NGO) declared "there is a congestion in Igdir as a result of trans border sex traffic in this city, the traditional roles of local women have become a lot harder with this new situation with the density of the traditional structure." This situation shows us that the new forms of economic life that the border creates produce similar experiences for the local women in the border regions, at least in the example of Turkey.

While talking about the example of Hopa, it is also necessary to touch upon the differences between the forms of Hemshin women's and Lazi women's experiences of the process. The condition of moving from the village to the city, together with the border, has been the case for Hemshin women. In the center, the Hemshin women who do not have any education and have only worked in tea and the field throughout their whole lives come face to face with the Lazi women who are more educated compared to themselves, "more from the city" or "socialites," as they call them. What is more, when the women who come from the other side of the border are added, these Hemshin women are left at the bottom of this hierarchy. Their spouses earning more money from the border commerce and the family economy being improved did not change the conditions of their a lot.

A reality that appeared with the border and can be generalized to Hopa or, more truthfully, to the Black Sea shore, is the fact that this condition the women find themselves in—the Black Sea women who can have what she wants, who is strong, and has a say stereotype—is broken down.

> You know, if the women were really able to make everyone do what they say and had a say in the family, could this prostitution come this much

inside our families? The man, you know, should either so respect or love or fear his spouse that he would not do this thing.

(FH_IW-3)

While being a woman in the Black Sea region is a situation full of hardships by itself, with the opening of the border, a new period started for the women from Hopa. They have to compete against the women who came from the other side of the border who are "educated," "groomed," "flirtatious," and "accessible by their husbands at all time." In this regard, the women are among the most affected by the chaotic processes that the border created in Hopa.

> Believe that these Hopa women of ours are very unlucky. Say I am among the lucky ones, but some are very unlucky. They especially do not come out of the tea houses. They don't help their women at their houses. Our women look after everything. Forgive me, but they work like donkeys, but there is still no return for it. I have heard from some of my friends, for example, they even threaten by saying, "It doesn't matter if you give it or not. There is a Russian in any case." The woman works. What will she do if she does not work? The woman is working for her children.
>
> (FL_IW-20)

Differences between the perceptions of the border by male and female subjects are striking. Unlike female interviewees, several of whom expressed their wish for the closing of the border, not even a single male interviewee expressed that. For the women, the border is a thing that disrupts the family structure. It causes their husbands to leave their homes and spend time with well-dressed women with whom they cannot compete. For men, it is a thing that disrupts the family structure but provides a very important economic income.

> I have never said, "Yeah, it has been good that the Sarp border gate opened, it is a new life." It is correct that a new life has started for us, but it has been very terrible for women, you see. Of the women and married couples, that is to say, the shattering of this family, that is to say, even those families that did not shatter have stood out of necessity. After all, the woman does not have any economic security. She cannot leave and go anywhere. There is a feeling of motherhood. She cannot leave her child and go.
>
> (FH_IW-3)

They are obliged to do the house work, look after the kids, harvest the tea, etc. They must continue the ordinary "womanhood duties" that have been stuck on them for centuries and also become prettier, look after themselves, and try to keep their husbands at home by satisfying them. What is more, the necessity of "keeping her man in her grasp" is normalized by both men and women. The statements of some of the women interviewees were as such.

Now our women are used to it. They have become accustomed to the sight of it. Another thing is they can't raise any protest either. They can't do anything, and what can they even do anyway? After all, we are also saying to them, claim your own spouses. After all, nobody is forcing anyone to go to anywhere. Isn't that true, though? Nobody forces anyone to go to anywhere. You should claim your spouses yourselves. You should also groom yourselves. For example, before the Georgians arrived, our women were not conscious a lot. Now they are a lot more conscious, and they groom themselves more.

(FH_IW-26)

In almost all the interviews that I conducted women whose husbands cheated on them with Russian women, they were dumped for them. However, I was only able to interview one of these women. What was told in other interviews was always either a neighbor's or a relative's story. Both with a feeling of shame and as a result of expressing it, bringing the necessity of doing something, the women told the "stories of others."

My downstairs neighbor is like that. She is a hairdresser. Her husband lives with a Russian for how many years. The woman is not very beautiful either, neither the man is with height and figure. He owns a night disco. The man found the Russian in just the place. However, this is a of group ten people or so. They always come and have their hair done by this hair dresser. The Russian who her husband dates comes and has her hair done by her.

(FL_IW-20)

The opening up of the Sarp border gate deteriorates at first the relationship of trust between women and men. The women do not know whether their husbands are going to come home, to another house, or to a hotel room with another woman.

My spouse has cheated on me. We have experienced the Russian event. This is not only one person cheating in this. The economy of the house-hold has collapsed. What should I know? There is no union left in the family. In a material sense too with this prostitution, it has also affected the material dimension within the family. That is to say, because the money that comes. For example, the woman is collecting her tea. If it is the man who takes the money, in any case, that money would go that night. The woman cannot even ask, and when she asks, she takes a beating and sits down to her place.

(FH_IW-3)

Before the border, the women did not have an equal relationship with their husbands. However, women being in the production process has provided them the right to intervene in some certain situations. The opening up of the border took away this, no matter how small it was, relatively equal space from them. Today, the lack of opportunities to work, and the fact that the tea does not retain any value in an economic sense, has imprisoned these women in their houses.

Relations with the extended family

I elaborated on the participation of ethnic identities in the commercial life and the conflicts that these relations created within the extended family in the previous section. Even though they participated in commerce with the support of the extended family, many families fractured while migrating to the city center and turned into nuclear families. As a result, the solidarity and control mechanisms that continued over the primary relationships in the village broke down too. If it needs to be reiterated, this situation is experienced in Hemshin families inevitably more intensely, because the process of migrating to the center is relevant for them. As this process started earlier in Lazi families and the decision-making processes operate within the nuclear family, its effects also remain at the level of the nuclear family.

According to Kandiyoti (2011, 30) the induction of villages into the national market economy has destroyed the traditional household economy. The wide family structures were exited, and everybody established their own household. Whether as the result of mechanization in agriculture or a result of an inclination toward market goods, a structural transformation was experienced in rural production, and this situation had a direct effect on the form of the household and its operation. The integration to the national economy that started with the cultivation of tea in the Black Sea region was also the beginning of the transformation in the rural structure of the region. Even though the cultivation of tea was done in rural regions, the fact that more money was brought into the household brought about the inclination toward market goods and the tendency to participate in the modernization processes. The tea cultivation that became the most important source of income in the totality of the region was the primary push factor for the rural-to-urban migration, especially for the Hemshins. However, the forsaking of the extended family structures did not happen immediately. The process was accelerated by entering the border economy. These new forms of relationships that were shaped with money "has teared up and done away with the emotional veil in the family relationship and has reduced it solely to a relation of money" (Marx and Engels 1998, 119–20).

While it was the eldest member of the family in the Hemshin families who was the decision maker before, this right has transferred to the man in the nuclear family. In many families, especially during the initial period of border's opening, a male figure who was present at home very sparsely but still was an authority was said to appear. According to Kandiyoti (2011, 31) the rural transformation that caused the economic basis in the traditional families to unravel also caused a great dissolution within the relations of authority inside the world of men by diminishing respect for the elders and causing the young married men to take the leadership role. According to her, the rural transformation did not change the inequality that already existed between genders, but it changed the social stratification between men. The changes related to women, on the other hand, were only supplementary to the changes that were taking place in the relationships in the world of men. The elder members in the household losing their right of say made the situation for the women in these families a little bit harder.

This dissolution that migration from village to the city created, especially the situation where the husband has the say as the leader of the household and not the father-in-law or the father, has caused new conditions of victimization for the women.

> There are those who are subject to violence at their homes. There is our neighbor, for example, we know that her spouse goes and beats her up. Why? Because she criticizes or because her kids are under hardship. She says, "You are not bringing home money"... As I am saying, if the man has three cents of money in his pocket, he does not plan anything related to his kids, but he eats and drinks handsomely that night and experiences his relationship with his woman at the hotel too. He can come home like nothing happened, I mean.
>
> (FH_IW-3)

The women are stuck between a family that says "you can divorce if you don't take your kids with you" on the one side and a husband who cheats and uses violence on the other. Their lack of any kind of power in the economic sense does not give them any means to create solutions.

> There are still those who live together because there are kids. That is to say, for their children. For example, for the family to not get disrupted because she does not have any place to hold on to. Where will she go if she lets go?
>
> (FL_IW-20)

The family of a man who had an affair with a woman who came from the other side wanted to prohibit this situation, but the "he is a guy, he will cheat" logic of the families has not changed what was experienced a lot.

> We had a neighbor... he was married with his uncle's daughter... this family put some pressure on. His older brother stopped seeing them. His mom drug him away. His brother did like this, but the guy does not come to the house, does not care for his kid, and does not care about his spouse. This time he slits the relationships in totality. He doesn't come home at all either. But the family does not have anything that they can do

after all. After all, what did this is "a man is a man, he can cheat" logic. In the end, he accepted or they accepted to be more truthful. After he came, the woman could not take it because of her pride. The woman took off and went, leaving her kid, three years old.

(FH_IW-3)

A man being seen as an authority in the family, and the acceptance of what he does as a "man's right" because of that, has deepened the desperation of women within these relationships.

It is bad, you see. She cannot leave. She has three children. How should she leave them and go? In our place here, you see the thing of "if you leave them and go, we won't give your kids. You won't be able to see your kids." She cannot leave and go. She bears it out of necessity... His family also knows, but they think, "He is a man, and he does that" and, you see, "Everyone who goes to the insides does that. What is wrong with that?"

(FH_IW-12)

The existence of prostitution has brought extra pressures on the life of the local women. The women are squeezed between not looking like the migrant sex worker and not letting their husbands be taken by them. How they are going to dress? How they are going to dye their hair? They are obliged to do it in a form exactly opposite of the woman coming from outside: the sex worker. They feel the pressure of being groomed and pretty like that at the same time.

Even the fathers are doing a thing like this to their daughters. They are establishing pressure to say, "Don't do those, don't wear these, don't do those." And for what? But I know there is a life like this on the outside, you see. When you dress up like this, you understand that it will immediately be perceived like this. He pressures more. Why? Because he knows of that disgusting filth. He both experiences it inside and is aware of it when you change a little bit. It could be your dress that you wear. Let's say it could be your hair or how you look. Oh, you do it like that? Be careful, you will draw attention. Don't dye your hair blond. But why? Those who dye are the Russian women, the streetwalkers. See, are you going to look like them?

(FH_IW-3)

According to Kandiyoti (1997, 7) arguing that the status of the women changes through their contribution to the production is a simple economistic approach. The fundamental factors that determine the position of women within the family in Turkey are extra-economic factors such as fertility, age, the immediacy of her position as a bride, or being a mother-in-law. This does not mean that family relations are independent from economy according to her, but the economic interaction with the local structures should not be disregarded. However, this "simple economistic" approach has been the fundamental factor in determining the position of the women within the family in our case. The woman from Hopa lost her right to say what she had, both in her relationship with her spouse and her continued relationship with the extended family. She could not join the economic activities that the border economy created, and the small amount of tea that she collected did not have an economic value. While these women harvested the tea, did the work in the field and garden, and took on all the household chores while looking after the kids were respected more still in the family, today the labor that they spend has turned into an "invisible" state. What is more, the fact that Georgian women worked in families where the economic conditions are good: jobs like taking care of the elderly and the children and cleaning the house. In addition, if the husband got together with other women outside, it caused the woman to turn into a meaningless object inside the home. These women made themselves with their "duties and responsibilities" within the household. The lack of any economic or social means to change this situation imprisoned the woman inside the home and rendered her bound to her husband.

The situation that appeared immediately after the reopening up of the border and that affected the women and the family life the most was setting up the men a new life with a woman who came from the other side of the border. This caused the increase in the divorce rate. However, a lot of women felt that they had to accept this situation because of the reasons of multi-layered, gendered division.

> We have experienced periods with a lot of shake ups. It is not that we haven't experienced them, but I would have been forced to take them all in. The only reason for this is never and ever. I will forgive my spouse in order to spare my child from having any stain on his name. For that I am a mother. Oh would I have gotten divorced if I did not have a child? I would never, ever stay for another second. I would have still gotten divorced even if I knew I would go hungry. I would not stay if I did not have a child.
>
> (FL_IW-29)

The dynamics that surfaced with the opening up of the border changed life and the perception of the youth for the generation that is recently growing up. At first, because the money being earned was all they thought of, this situation was not noticed at all. The prostitution sector, the families who were shattered, the double marriages, and now the drug problem that is being spread out among the youth has brought the problem forward.

> Now the kids are experiencing so many problems. They are so much more than what exists in other regions. There are things that they learn before their time. They have seen the things that they should not have

seen before their time. Generations we have said are not educated like that, and these ones are getting educated way too fast. They meet with certain things very fast... The drugs come in from the border very easily. Our friends and the kids of our friends who we are close with have gone into treatment. The kids of our friends who have political culture were left without controls and went under treatment. They used pills. These are things that the border brought. Those who were caught with the biggest drugs were from here, from this gate. It is still going on too.

(ML_IW-10)

The kids and the youth experience the deformation that the woman experiences in her relationships with her spouse and social environment both in their relationships with their parents and in their social environment. Especially during the initial times that the border was first opened, the economic conditions created opportunities for the youth to earn money. These youth, who started earning money by carrying stuff with handbarrows, are participating in some way with this economic life today by waiting tables in the entertainment sector.

When we were kids, we climbed on top of the cars and took copper. There was a Kurdish Aladdin here. We used to go and sell it to him. We were doing this when we were kids. Look at the relation that we engaged. I know that I have jumped to the other side of the bridge because I was going to get beaten up by someone. I mean, I was a kid of 11 or 12 after all. We have climbed up on top of the car. There the drivers and such have seen. They said, "Halt. What are you doing?" We slid from there and went, we felt the need to hide. Think about it: suddenly a criminal. You enter a relation of crime. I mean, how could this perception not change you? For example, in the most recent times, it is said that if the kid cannot get an education, let him do waiting at a restaurant and earn money. It is not important what they do there. The spreading of the abuse of drugs is being fed from there.

(MH_IW-39)

The recently increasing smuggling of diesel, cigarettes, marijuana, and heroin increased crime as a result of this bringing a "Mafiazation," and the spread of usage of these materials among the school kids come up as an important problem. The fact that the young get inclined toward this type of smuggling business in order to earn money, or set becoming a taxi cab driver as an enviable goal and not pursuing a university-level education, is mentioned as the disadvantage of the economic life created by the border.

In the end, we are not happy with this gate, we as the people of the region. Marijuana, heroin—all the filth has entered. Our youth have lost their boundaries. That are to say, we thought of it in big cities. Even I have seen around our place the cars that were parked on the side of the streets. Marijuana has started to be smoked like cigarettes.

(MH_IW-11)

The feelings of trust in the youth and the children in these families are shattered, and their nervousness is rising. This is the summary of one of our interviewees.

> How should I say? The concept of family has ended. Not taking responsibility started in the families. I mean, let's say in the household leaders, the fathers. You see when one father does not come home, when he comes at 1:00, 2:00, you think about the condition of that kid. The kid gets affected either he wants to or not. I am making this up. He took the money of the tea and spent it on in a night at Hopa in restaurants and discos and then with a woman... It hasn't been good. That is to say, our village life has ended.
>
> (MH_IW-11)

Women imprisoned at home

The prostitution that intensified with the opening up of the border took the social spaces that women used outside of the house from their hands. It is to such extent that the women have been rendered to a condition where they can't even satisfy their most fundamental social necessities outside of the house. The women either lost the use of spaces, including first of all the streets and places like movie theaters and cafes, or had to limit the use of them within certain time frames.

> There is no place for entertainment, and the ones that are there are the places for Russians. We cannot go out in the evening and walk around anywhere. You cannot tour around outside with ease, because there are Russian affairs. There was no such thing before them... You can go down there up until a certain hour, but you cannot go down after a certain hour. For example, let me walk with my daughter at a late hour. Let us go to the market place. Things like that do not happen. It is for sure that something will be stuck. They will say something, or they will think that you are Russians. The life is hard in that regard. It is hard for the women.
>
> (FL_IW-21)

Like one of the interviewees said, her older brother and father did not let her go out before and then her husband. Now no streets that she wants to go out in are left for her. And in the manner that it was expressed by someone else, "Hopa is a small place. There are three streets. You cannot use one of them during the daytime, too, and none of them after eight in the evening" (FL_IW-29). She cannot enter Texas Avenue in the daytime. Her getting out of the house after 8:00 in the evening is not appreciated, and she really is disturbed. She cannot even get out with her spouse. She cannot stroll around the seashore all by herself. If she does, there are conversations about her.

> This place seemed like it was good with the opening up of the border, but I think that it was good. We normally used to go out more comfortably before, but now we can't. If you get your hair dyed differently or wear black, immediately people look at you with the eyes of "Are you a Russian or Georgian?"
>
> (FH_IW-12)

As such, the public life of women that was already restricted was put under some sort of siege in practicality. In addition to all these, the almost non-existence of job opportunities for women imprisoned them in their houses even more.

> Living in Hopa as a woman? That is to say, it is difficult after 8:00. It is not such a huge problem during the daytime. However, we do not have a probability of going out and strolling around at the seashore. Does a person go out at that hour? It is like because there are Russians, people who drink alcohol and the like. Going out in the evening and sitting at a park with a tea service, strolling around at the seashore with the family. There is no situation like this. I mean, you go back to your house after 8:00. It is as if after 8:00 is an hour that the prostitution takes place.
>
> (FH_IW-4)

The women of towns live a more closed life than the women of rural regions, but behind this closedness there exists a richer social life according to Kandiyoti (2011, 35). While this condition holds true for a woman who has spent the whole or a large part of her life in the town, it remains inadequate to explain the condition of the women who had to migrate from a village to the town. At least the discourse of the women who moved from a village to the center (that we have interviewed in this study) was in the direction of how their social lives were restricted in the town. Here it is also necessary to consider the extra restrictions that having to live in a border town created like the use of space. As one of the simplest points, the women who could have used all the spaces in the village are in a condition where they cannot use one avenue, Teksas Avenue, at all. Other places and streets, on the other hand, could not be used after 8:00.

As a result, the life of the woman from Hopa is in a condition where it is squeezed more into the household with the opening up of the border. There is almost nothing else that they can do besides organizing house gatherings (*gün*) and visiting each other's houses. What is more, the connection of many women who live in the urban center of Hopa with the village has been severed except for a week or two during the summer. As a result, Hopa was a place where the social life was more active and where the women used to be freer in the social activities compared to many places of settlement in Turkey as a result of its political history. It turned into a place where it was "not possible to have fun" especially for the young and middle-aged women. To say it otherwise, a problem of "inactivity" among the women in Hopa has formed.

> You clean the house and prepare the food. Then it is already 12:00. Between 12:00 and 3:00, I hang out some place, with a neighbor or at a friend's, and come back. I sit at home, and the kids arrive... We also do home gatherings (*gün*). That is to say, there is nothing to do in Hopa. You are either going to work in the fields, or you stay at home and cook for your husband. That is all. There is nothing else to it. You go and do a little bit of gossip from time to time. You can unload there a little bit, and that is all there is, you see.
>
> (FH_IW-7)

Victim or beneficiary: The complexity of winning and losing

In order to understand the ways in which women are adapting to the new dynamics that have surfaced following the opening up of the border in Hopa and elaborate upon the new vulnerabilities that have appeared, it is useful to look at Kandiyoti's concept of "patriarchal bargain." According to Kadiyoti (1988, 285),

> Like in all systems of hegemony, the systems of male hegemony have both protective and repressive elements and in every system women also have their own sources of power and autonomy. As a result women can be as attached to the systems that seem like they are oppressing them as much as men. However the "patriarchal bargains" depend on the assumption that certain mutual expectations will be satisfied and the quality of these expectations may change from society to society.

Kandiyoti (2011, 126) suggested that "in any given society that can display varieties related with class, cast and ethnic origin, the woman build their life strategies within the framework of a series of concrete necessities that originate from the system that they exist within and" has deemed "the concept of patriarchal bargain" appropriate "for these." According to her, the patriarchal bargain

> [p]oints out to the existence of a series of rules that regulate the gender relations which both genders agree and consent to but nonetheless one that can be resisted, redefined and reviewed. These patriarchal bargains, make a strong affect in determining both the subjectivities of women and the quality of the gender ideology in different contexts. At the same time they affect both the actual and potential forms of active or passive

resistance of women. Most importantly, patriarchal bargains are not ahistorical or fixed; they are open for the renegotiation of gender relations or to the historical transformations that open up new spaces for struggle. 'The passive resistance' of women in these patriarchal relations that are changing 'takes the form of demanding their own share within the framework of this patriarchal bargain: in exchange for protection and security; obedience, compliance and acceptance that the honor of men is in fact related to her own respectable behaviour.

(Kandiyoti 2011, 139)

The bargain of the women from Hopa who are imprisoned within their houses has been the exchange for their husband earning enough money to sustain the house, consenting his entrance to any form of trade that the border economy creates, and accepting the results of this.

The rural transformation in the Black Sea region, mentioned before, deepened the stratification among men as Kandiyoti said. This process of transition gained another dimension with the opening up of the border and at the same time deepened the inequalities that existed between women and men. The existence of women started to shape according to the form of male participation in the conditions that the border trade created. If a woman was also a Hemshin, these inequalities deepened even more at this point where economy, gender, and ethnicity intersected. She was deprived of the right to say what she obtained by participating in the production process or taking on the totality of this process in her rural life by being imprisoned within a house in the city.

Even though the dissolution of the traditional patriarchal system based on the land had the potential for emancipating women, according to Kandiyoti (2011, 50), this remains only a potential: "When women get separated from their traditional duties and start to have free time, rather than entering the society as productive members," they turn into "a symbol of prestige for men through conspicuous consumption." What is more, when women who are "well groomed, well dressed, and young" who come from the other side of the border are present in the location as sex workers, like in our case, the situation gets more difficult, and a context of unequal competition appears. In this context, it is almost impossible to become a "symbol of prestige" for the men. She is required to redefine her own womanhood in accordance with these migrant women and reconstruct it according to that. She is forced to keep her husband by getting dressed more beautifully, looking more beautiful, responding more to her husband's sexual needs, or consenting to share her husband with other women.

In their research on the women who do house work, Kalaycıoğlu and Tılıç (1998, 235) say that when handling the condition of women in work life, evaluating it only within hegemonic relations of capitalism or patriarchy generally leads to discounting women as individuals. According to Kalaycıoğlu and Tılıç,

[I]t is the fact that the events, relationships and the context, she determines strategies that are appropriate for her own condition and that will make herself happier by remolding all the ideological affects in her own pot. What is more, this is a legitimation process that the individual experiences continuously, every day, every moment. This legitimation process, is the major factor in determining the position of woman in family and the society.

In order to live with the knowledge that her husband gets together with another woman every day and to accept this, the woman from Hopa also needs a legitimization. While she is doing this, sometimes by turning a blind eye, sometimes she legitimizes it by saying that she stays for her kids even despite knowing it. These discourses are the coping mechanism for the women with the conditions that the economic life on the border creates and a bargain that she makes with the patriarchy. On the one hand, by migrating from the village to the city, she gets rid of a lot of the hard work in the village and starts to live in better economic conditions. However, for this she gives up on her womanhood. A lot of the time, she legitimizes her existence in the household by sanctifying motherhood.

As Erman (1998, 211) also states, the rural-to-urban migration is a phenomenon that is desired by the rural women, and it carries great expectations: a promise of a better, more comfortable life for them. The city for these women is

> [a] place where she can get away from the repression of her husband's family and relatives, establish her own family comprised of her children and husband, get rid of the hard work of the village, working like a "slave" both in the house and on the fields, her husbands "machismo" can soften up, what's more where her children will find the opportunity for education, the family can benefit from the health and other services of the city.

The Hemshin women who migrated from village to district center were also freed of a village life where they had to live together with their mother-in laws and father-in-laws, work in the tea gardens and the fields, and do household chores.

> Back then, we would get up when the call for prayer was sung and prepare [breakfast]. Do you have what we used to have then? There was a headscarf, like this. We used to cover it like that. They cautioned me when I got married. My aunt and the wife of my uncle tied a headscarf on my head. I didn't use to cover my head. They said, "Let's cover your head that morning." It was almost like I became a bride that morning. If you worked until the evening, you weren't able to speak, you weren't able to drink water in their presence. You would work and work but not be able to drink water... It was like that, you see. The girls today live very good,

and it is very good of them. They don't get oppressed, but we were oppressed a lot.

(FH_FG-1)

However, the fact that women did not migrate to the big city but to the district center restricted their use of space. On top of that, they had to deal with the problems that were created by being in the border district (which we mentioned above). Perhaps because of these problems that they faced in the center, they frequently voiced the longing they had for village life.

The old was more beautiful. There were festivities, there were girls, there were brides. Now brides have all gone to where their husband is, and the kids have gone after them too. Now if there is a house here, it is always a wife and husband. There is no more bride or such any more. Everybody has gone. The villages have become vacant. There used to be old ladies, brides, grandchildren.

(FH_IW-28)

While the process of rural-to-urban migration contained positive and negative sides within itself, and while those who migrated experienced this process in its normal course, the existence of the border and the new conditions that it created intervened in this normal course. It added new dimensions to the positive and negative aspects. This created new experiences in the lives of both the Hemshin women who migrated to city from the village, the Lazi women who settled in central Hopa, and the migrant women who came from the other side of the border. No one completely lost or won. The women especially obtained material gains that would especially ease up their lives. Women also benefitted from the economic opportunities that the border created, even if it was through their spouses, in exchange for those things that they sacrificed for their children or in order to continue the unity of the family. On the other hand, they were subject to new restrictions and status losses in their social lives. On the subject of these losses and the inequalities that they faced, there is no search for a solution other than certain individual interventions.

To sum up, as discussed above, with the border, women subjected to new victimhood "bargained with the patriarchy," shut their eyes to their husbands' affairs with other women in return for livelihood and the opportunities provided by the city, and "justified" this situation with the blessing of motherhood.

The "otherness" of the migrant women

Women coming from the other side of the border for daily trade, temporary work, long-term job opportunities, or to settle, whatever their motivation for coming, remained in mechanisms of exploitation and oppression created by the patriarchal system on both sides of the border. Whereas local women experienced oppression and the disregard brought about by isolation, immigrant women in contrast experienced them by being drawn into the trade mechanisms operating on the border. When we look into the reasons that push a woman to such a journey knowing full well that her labor, her body, and her identity will be exploited, the only answer that can be given is poverty. These women who were well educated and had professions in their own countries found a way to get rid of the horrible poverty that Georgia fell into after its declaration of independence. They crossed the border, whatever the cost, to get involved in economic life that emerged on this side of the border. However, the trade, often carried over the bodies of these women, did not only commodify the women but also brought about sexist and nationalist discourse in the region. In her study on Chinese perception of Vietnamese women in Hekou, the small town at the Chinese border with Vietnam, Caroline Grillot (2012) examines how "the other" is constructed through the body of women in this border town and the ways in which stereotypes and narratives continue. According to her (2012, 144),

> Vietnamese women who cross the border and now live in areas bordering China have quickly come to represent their country—in the eyes of Chinese people—as a figure of marginal femininity… Contradictory and accommodating images emerge and then expand in the discourses, portraying these women as submissive spouses, tireless workers, prostitutes, manipulators, heartless pragmatists, devoted companions, and ambiguous merchants.

Such a perception of Vietnamese women is part of the dominant view that emerged after the opening of the borders in the post-Soviet period. Women who were an important aspect of small-scale border trade stimulated nationality discourse. In relation to the small-scale trade within border regions, Archer and Racz's recent study (2012, 70) is a good example of showing the ways in which gender is determinant of economic structures and how women play an important role in cross-border trade.

> Smuggling as a social practice both supported and subverted traditional gender roles: During the bombing of Serbia in 1999, as men were prohibited from leaving the country, women would cross the borders in their place to buy goods and sell them back home. Yet, larger-scale illegal smuggling has remained reserved for men… Also, the type of items traded by women were characteristically associated with the domestic sphere assigned to women, taking for granted that they are better at choosing the type of food, clothes etc. that is most demanded and profitable in Serbia.

As in the above case, although women become an important component of the trade at the border, they continued to be kept in a frame defined by traditional roles of womanhood. We can see women who are actors of economic life both in Hopa and on the other side of the border only in small-scale trade in housework or as heavy workers in the entertainment sector. Moreover, it is these women who are subject to legal sanctions brought about by almost daily crossings of the border and by illicit work and prostitution. During the period I was in the field doing interviews, there was an operation, and a majority of migrant women were deported. The others were hiding in their houses. Even though I tried to reach them through one of the beauty shops that they frequented, I was not successful. Clear statistical data on the number of these women or the frequency of their entrance and leaving cannot be obtained. This is because these women work illegally, and their visa status forces them to leave the country periodically and come back.

The reopening of the Turkey-Georgia border caused many more problems for the Georgians, especially for women. Dudwick (2002; cited in Dursun 2007) states,

> [W]ith the beginning of 1990s, the drop in family incomes, together with reduction of state-support and the introduction of fees, with the break-down of health services means that poor children and youth are system-atically excluded from opportunities that will allow them to compete on an equal basis with others. As the outcome of these, an increasing number of poor children had stopped going to school; many worked informally with their parents, while others worked independently as traders, goods handlers, or assistants, some doing heavy manual labour at young ages.

Considering their involvement to border trade, it is possible to talk about three different types of women who came by crossing the border and parti-cipated in the economic life during the time that passed from the opening of the border until today. It is also possible to talk about ones who settled down on that side of border by means of marriage and suffered heavy gender inequalities.

Shuttle traders

The women who came as shuttle traders during the initial period, who passed the border with the amount of goods that they could carry, selling them and buying goods from Hopa, entered the economic life of the town both as "sellers" and "buyers." These women started to come with their families to sell the goods that they had in their houses. Eventually this turned into a part of suitcase trade that started at the Black Sea shore and slipped toward Istanbul Laleli.

For the women on both sides of the border, the opening up of the Sarp border gate caused a serious change in their lives. The women on the other side of the border came to Hopa to sell whatever they had in their possession in the Russian markets when the border first opened. These consisted more of daily crossings. According to Dudwick (2002; cited in Dursun, 2007),

People tended to sell possessions in three stages; beginning with personal property such as jewellery received as wedding gifts, linen, clothing, or fine crystal; then furniture, appliances, and cars; and finally their homes. Sometimes Tbilisi women exchanged used clothing, costume jewellery, or perfume. For respondents with nothing left to sell, their own blood provided the final source of income.

In that process, women became an important figure of the cross-border trade. They made themselves available at any time and in most places in the city center. In the study of Heckmann and Aivazishvili (2012, 195) in which they give the recount of two women who crossed the Azerbaijan-Georgia border every day and tried to sell the goods that they had at hand, what these women say

> can be interpreted at the common sense level as petty-traders engaging in this trade because they lack other income and resources. Leila argues that she needs to earn her living in this way as she has to look after her children and make ends meet.

For them, Dudwick (2002, 195) explains the situation of why these women became the main figure of small scale cross-border trade as such.

> Although women have been harder hit by unemployment than men, their secondary position in the labour market has paradoxically made them more flexible and adaptable. Women, perhaps because they had multiple identities as workers, wives, and mothers, were able to adapt more success-fully to the loss of formal employment than did men, whose social identity was more tightly bound to their role as breadwinner. They sought service sector jobs locally and abroad as housekeepers, nannies, and waitresses. Women had come to play an important role in trade, even travelling abroad without their husbands and absenting themselves from their families, behaviour once considered unseemly.

On both sides of the Turkish-Georgian border, the victims of the chaos created by the border were women. On the Georgian side of the border, women who chose to leave home and cross the borders to go in other countries, especially Turkey, for financial reasons were exposed to countless humiliations and dangers that they had to deal with on their own.

Some women travelled in small groups by bus, train, or air to the Russian Federation, Turkey, Hungary, Poland, and other Eastern European countries as often as once a month. Others traded inside Georgia. The wives of unem-ployed miners in Tkibuli came to Tbilisi, where they shared small rented rooms, to trade in produce. Their concern for the daily welfare of their children and family was a strong incentive to move into such unprestigious activities as street trade. The practice of returning from organized tourist trips abroad with suitcases full of scarce, foreign-produced commodities to sell expanded rapidly into an international "shuttle trade." Women, in particular, moved into this arena, ignoring the low social estimation of commerce as "speculation." Some engaged in prostitution locally in new urban brothels or abroad. Female traders had to overcome problems with police and organized crime (Dudwick 2002, cited in Dursun 2007).

After a while, when they had nothing left and the Russian markets closed, the entertainment sector surpassed all commercial activities. Thus, a lot of women started to come to Hopa as sex workers.

Cross-border marriages

One of the ways of coping with the deep poverty created by the dissolution of the Soviet Union and the neo-liberal policies of the newly emerged countries for the women in these countries has been going into transborder or interborder marriages. At the Turkey-Georgia border, too, one of the methods employed by Georgian women to cope with poverty has been to "settle by marriage."

It is necessary to include both those who received citizenship by arranging fake marriages and those who came to really marry. Many of those who had fake marriages and received Turkish citizenship in Hopa could not stay in Hopa and settled in big cities in order to work.

According to the information that the population administer gave, the number of those who came through marriage started to decline after 2004, because until that time, those who married could become citizens in a day. For this reason, during that period there were a lot of fake marriages according to the census bureau representative. He said that after a three-year residency requirement was put in place, the ratio of marriages with women who came from the other side of the border declined.

The women who had a real marriage, had children, and settled down experienced a lot of hardship because of the legal procedures or cultural differences. A Georgian female interviewee complained that she could not get her ID card even after 15 years of marriage. The words of a Georgian woman who married and settled down in Hopa show the difficulties that she faced.

> I had a lot of difficulties for two to three years. I was, of course, comparing. I was living this, and it happened different here. But I still got used to it. The difference is very huge indeed. Now, how should I say, I can't remember exactly, but I have experienced difficulties. I told myself, but I did not make it evident to anyone. Now there is something. For example, if it is not a very close relative, I don't go to weddings. I cannot adapt now. There is Horon (the rhythmic, folkloric dance—translator's note). There is something. You sit down. How should I know? I stay away a little because of that.
>
> (FG_IW-9)

There were also various cultural difficulties in the process of getting married. The members, especially the older ones, in the families in which they came into through marriage tried to claim them or tried to have the society accept them by giving them Turkish names. "My mother-in-law wanted Inci, for example. I did not want it." What is more, a lot of them had to change their religion and become Muslim in order for the family to accept them.

Even when crossing the border and settling occurred through marriage, the fact that a woman initially crossed illegally could lead to consequences such as not being able to acquire citizenship for a certain period of time, joining the family of the man and becoming separated from her own family, and changes in identity and citizenship. That there was not a home to go back to increased their victimization. According to Palriwala and Uberio (cited in Williams and Yu 2006, 59–60), this type of marriage always had three effects on the women: The woman lost her rights of inheritance in the house she moved out of, she lost her power to negotiate as she did not have a say in the new house, and she found herself in a difficult situation if the marriage resulted in divorce. Indeed, the story that a Georgian woman I interviewed told me about her cousin confirmed Palriwala and Uberio. This woman married and settled in Hopa some time after the border was opened; she had problems with both her husband and his family because she did not want to do housework or work in tea picking, and she had to live with the family of her husband and so did not have the opportunity to set up a home for herself. This woman, a music teacher by profession, could not work outside the house either, as her husband did not permit it, and she finally divorced her husband and left Hopa. However, she did not settle in Georgia with her family but in Istanbul to work. One reason for not going back to her family was that they opposed to her marriage with a Turkish man, and another was that she was worried that she would not be able to find employment in Georgia where economic conditions remained poor.

What is more, these women experienced difficulties similar to those experienced by the local women of Hopa, and in general, they continued a life that was imprisoned within the household, devoid of opportunities for work. Experiences of those women who settled through marriage can be defined as "structural intersectionality." Crenshaw (1991, cited in Acar *et al.* 2008, 3) defines two different type of intersectionality: political and structural. Structural intersectionality occurs when inequalities and their intersections are directly relevant to the experiences of people in a society. Accordingly, an immigrant woman's experience of domestic violence might differ from a native woman's. The former might face multiple oppressive mechanisms or discriminatory practices hat harden her experience and capacity to deal with the experience of domestic violence.

Transborder marriages happened in two main ways. The first was the transborder marriage of individuals with different ethnicities or cultures. The second occurred through relations such as kinship within the same ethnic connections or through mediating persons, and the main factor defining work preference is ethnic connection and continuity (Yıldırım 2015). While the first type of marriages were common in the period immediately after the opening of the border in Hopa, they diminished in time as a result of complicating regulations concerning the right of citizenship. Today, the second type of marriages are more common, where Hemshins take wives from Hemshins living inland areas on the other side of the border, as I described in the previous section.

Sex workers

A large amount of gender discussion on border studies is carried out about the entertainment sector and women who are included in this sector as sex workers. This situation is also relevant for our example.

At the beginning, like all other sectors that were not settled/institutionalized, an environment of chaos reigned in the sex industry. Negotiations were made on the street and in abandoned houses, and even cabins near the tea fields were used for prostitution. Women did not have security of life, and they were subjected to widespread violence. There were many cases of violence against these women such as abduction, rape, and beating.

> There were those who died. And there were also suicides that we could not guess what it was about. They were thrown out from the hotel balconies two to three times. I mean, what kind of thing they experienced, we cannot know that either, but after all, these are the things that actually happened. There have also been ones who were murdered through violence, and most of the time, they are unresolved cases.
>
> (ML_IW-10)

With the drawback of prostitution from the streets during the day time and establishment of certain rules, albeit informal, for its operation, it became easier for it to be normalized and accepted.

> If it is a lady who comes from there, they had an image like this. It is certain that she was seen as a streetwalker in this way of life. They were looked at like that. At first, in the streets, everyone and I looked like that too. Now these businesses have changed compared to the older times. Now nobody cares about it. It seems to me like they have gotten accustomed to it or that the initial hunger ran out. Now that kind of life starts after 11:00 or 12:00 at night.
>
> (MH_IW-1)

Today, the entertainment sector which is said to keep Hopa on its feet, continues thanks to these women. As we have mentioned before, even though we could not obtain any information about their numbers, how they came into the country, or what kind of network they had here, it is possible to reach some evaluations from what some of our interviewees said.

Now I can't have much interaction with the women who work at the hotel anyway, because they don't have much business with me in any case either. In general, they have a leader. Two ladies. She comes, for example, buys and goes, and that's it. When I say leader, I mean the women who bring them here, you see, the women who market them... I don't know if this exists in every hotel, but I know one or two people who work under their orders. There is one, for example. She rented a house with three to five rooms. She feeds and looks after the girls there too. She finds clients for herself and takes money from them.

(FH_IW-31)

These women have a strategic partnership with the owners of the places (hotels, restaurants, or bars) no matter if they come and work on their own or if they come through the mediation of a network. It is such that these women need a place in order to earn money, meet their clients, and work in an environment that is relatively safe. The owners of the places, on the other hand, are earning money from the women who work illegally.

Now the pimp of Hopa, the first pimp is... but they have not done pimping in the sense of pimping. They have done it in the sense of a space. It is a cleaner thing. They opened the hotel and said, "Come and do your business, and give me your money." They said, "Take a shower, have a drink, pay the money for the drink, use my bed, do your business, and go."

(MH_IW-39)

The guy brings it himself. For example, I am a hotel owner, or whatever else, I am in that hotel. Now I go and bring the women. The women stay at this hotel. After all, the guy earns money over that.

(MH_IW-3)

The situation of "outlet" that appears in the ethnicized division of work life is also related to the sex industry in Hopa. The women who are younger, more beautiful, and will earn a lot of money are sent or taken to big cities when they cross the border. The women who stay in Hopa are, on the other hand, those who are older, who have families and kids in their own country that they have to look after, and are cheaper—by an interviewee's definition, "with defects" or "left on hand." The description of a shopkeeper in this street was this.

The old people, those who are 45–50 years old, come too. The young segment comes too. There are those who are 70, and that is because everybody has their client according to their own level. That is the reason for it... There are generally Azerbaijanis and Georgians here. We do not come across Russian much. Here is like the first stop in any case.

(FH_IW-26)

There are no problems as long as the shopkeepers benefit from the border just as it is with the shopkeepers on the street. They earn money from the business that is worked around here and are happy with the existence of the border.

> There are beauty shops in Hopa in a serious sense. Now the women are groomed. It is marketing, after all. She will market herself, that is to say. After all, these women go to the beauty shops every day. That is to say, this is their income. They do shopping for clothes, hairdressing, and the look, etc.
>
> (FH_IW-3)

Tverdova (2011)'s summary shows very clearly show how and why, after the collapse of the Soviet Union, the entertainment sector and sex workers showed up. "The economic shock that hit the former communist countries resulted in the immediate deterioration of the living standard among the general population" (Ashwin 2006; Lavigne 2007). Women, however, suffered more severe economic consequences than men. Unemployment, nonexistent under Communism, was much higher in the female labor force, and women had more difficulty transferring from the public to the private sector due to the widespread social stigma (Bridger et al. 1995; Kay 2006; Ashwin 2006; Shvedova 2009). Those able to keep their jobs experienced terrible wage delays that lasted for months and, in some cases, even years (Ashwin 2006). Unimaginable hyperinflation in the 1990s made money worth less by the day, and the bankrupt government was unable to provide a safety net.

Domestic workers

Recently a new phenomenon appeared: women crossing the border for domestic labor, i.e., cleaning and elderly/child care. The number of women working in the houses of Hopa is more than a few. While some are hired as tutors for music classes for children, many are employed in cleaning and elderly care, and many of them are living within these houses. This has been a situation that enabled for the softening of the perceptions of these women. Those who came from the other side of the border are not that much of an "alien" for the people of Hopa anymore.

> There are a lot of those who came as care workers for the elderly. There was one in our district too. My aunt was ill. She was coming and looking after her, and she was such a good person too. The way she spoke, the way she would sit down—they were different in a cultural sense.
>
> (FL_IW-25)

The changing global economy created what Maria de la Luz Ibarra (2000, cited in Mendoza 2009) termed the "new domestic labor" in which the structure of social reproduction was transformed. In the past, employing a paid household worker was a luxury afforded solely by the upper class. However, due to the availability of low-wage immigrant labor, many middle-class households could afford housecleaning, full-time nannies, or elderly care workers. Women working in this occupation are also more diverse than in the past with many educated women fleeing countries with few employment opportunities to work in domestic service.

> There are also ladies who do dishwashing at the restaurants, cleaning at the hotels here, but generally, you know, let me say 90 percent of them come for prostitution. My spouse also has a hotel, but my spouse does not operate it. He rented it out. There are cleaning ladies there. They do the cleaning up of the hotel, but after all, they stay at that hotel, but they prostitute themselves on the one hand. Those who come, let's say there are 10 ladies, two out of that clean. They have really come here to look after their own families, but eight of them came here both to look after their family, you know, as I said, she seems like a cleaning worker but also does prostitution too like the rest. That is what I mean.
>
> (FH_IW-3)

The way in which a man from Hopa praises the border is in fact an affirmation of women's social status and men's expectations and the double burden of women.

> However, today everybody in every household who has an elderly person, a Russian lady and a Georgian lady work. I mean, it is our ladies who had taken this into their own houses. They take them in their own houses in order to look after the elderly, and then they have them sleep at night. There is one even in my home. Then why are they complaining? Then let them sort those jobs themselves, and let Russians not come. It is our ladies who work. The Russians, they have them work at home. They make them look after their mothers and fathers. They work them at the tea work or in cleaning duties. In these house duties, you know, it is the ladies who take care of that in Turkey. It is also the ladies who take her in this house. Then why are they complaining? You are both inviting and make them work and complain about it. The one that you have working in the house is good, and the one who strolls around on the street is bad?
>
> (ML_IW-19)

Mendoza examines the temporary migration of Mexican women who reside in Mexican border cities and travel daily to American border cities to labor in domestic service. He observes that (2009, 24) gender, race, ethnicity, class, and nationality work together in shaping the experiences and opportunities of women working in domestic service. Because of the prevailing gender ideologies regarding domesticity, women have typically been the predominant workers in this sector. This is inevitably the case in Hopa, since these gender ideologies have powerful effects on both side of border, as we saw in this chapter.

Local perceptions of migrant women

Local male perceptions of migrant women

Özgen (2006), in her article where she focuses on sex workers in Igdir, names this phenomenon "transborder prostitution." In Ozgen's interviews, men and women who come from across the border described this situation "with words filled with a bitter mockery, disdain and scorn" that is "towards ownership of the women." The women involved in prostitution are always "the women of the other" (2006, 126). For the Kurds, these women are Turkish. For Turkomans and Azerbaijanis, they are Russian or Dersimite. That is to say, these women are scorned by referencing their nationality. In Hopa, however, even though some similar emphases were made by some of our interviewees, for many interviewees, especially the men, the reference point has not been nationality. What is more, it is very clear that the men from Hopa, in contrast to those in Igdir, respect these women and see them as superior to their own women in their education, looks, and how they dress.

> Their women are much more cultured than our women. Let me say that from the beginning. They buy bread from here, and I still don't know one of them who leaves without thanking. "Thank you." All of them, and ours say, "Give the bread" and "fill it in the bag." What I mean is that their level of culture is higher than ours. Oh, okay, maybe they have fallen into that business, but we don't know how they have fallen. Most of them are highly educated, I mean. Don't regard that they are that thing, I mean. Maybe they are doing this business, but most of them are cultured people. What do I say? They buy bread. I have still not known a single one of them, of my clients, who leaves without thanking. In general, it is always the lady clients who come here. For example, they buy bread, and they always speak very kindly. I mean, "thank you" and "we are glad." For example, there are ladies who stay in the above floor, and they sell goods. We are like friends, I mean. There is nothing that says they should always be looked at in that regard, I mean. And I don't know if she earns her keep, that way I mean.
>
> (ML_IW-8)

The approach toward these people generally provides a mixture of feelings of uneasiness and admiration.

> With the border, a Natasha phenomenon has entered into Turkey. If you ask me, culture wise, the Black Sea's culture has changed. They were kinder people even though they were called Natasha. All of them consisted of

people who were educated. Culture wise, they are more cultured than us. I don't talk these about the Georgians, I talk these about white Russians. You see, with their clothing, the way they stroll, the way they sit down and stand up, there had been issues that they constituted an example... There is no such thing as no. You see, the Natasha came and disrupted the culture. The people who come, they all know a foreign language besides their own language. On the other hand, knowledge of foreign language in Turkey is almost nonexistent.

(ML_IW-19)

Many of those ladies are university graduates, with their culture and the way they dress up... But how was it especially our ladies who were living at the Black Sea in the past. Our ladies do all the work of the house, including me, preparing the food, harvesting the tea—though we do not have tea, however. The ladies harvest the tea, ladies fire the stove, ladies, excuse me for saying, look after the cattle, ladies are tired ladies. There is no time for makeup, dressing up, or putting on perfume. He saw a pretty lady, a lady with perfume, and he went after her.

(ML_IW-19)

The discourse about the women of the Black Sea region was expressed by everyone: they did not care for themselves, gained weight by indulging themselves with food, bearing children, and then became uglier, but after seeing the women who came from outside, they started to care more for their appearance in order to draw their husbands back home. A male interviewee summarized this cruel point of view as such.

Our spouses have changed too. It does not work with only eating and eating and getting bigger at home. Our spouses have become more caring toward their husbands. They started to become more beautiful in order not to let them get taken away. Of course, there is also this. You see, this has become a kind of competition. That is to say, there is no one eating kale and sitting in the corner saying, "Let me plant two corns and collect tea." There is no life like that. I have a kid. My husband would not throw me out. That does not exist either. That has also finished.

(MH_IW-2)

A critical approach toward the economic conditions that force these women to prostitution can be felt in the interviews. This does not necessarily correspond to a formation of political consciousness. These women commonly are not perceived as that "bad." The common discourse of almost everybody was in the form of "They are human too" and "There is so much poverty on the other side, and they are doing this job out of necessity." This pseudo-humanist discourse, on the other hand, is no different from naming the prostitutes in brothels victims of ill fortune (*kader kurbanları*).

I have seen people who were crying, well educated, selling their bodies for money. She said it herself, "I feel so torn apart." She is very cultured, educated, but there is no money in the country she lives in. I have witnessed a lot that they cried to anyone who treated them with a little bit of care. They don't do it because they want to.

(ML_IW-10)

Despite male interviewees acknowledging the economic difficulties that force these women to prostitution, they stress not labor or class but "civilization." This perhaps is no surprise for Turkey where, throughout the republican history, modernization and Westernization have been the dominant discourse. An age-old clash between modernization and tradition in the minds of the locals creates this awkward desire for Western culture.

Now we here, they are called Russian ladies, but go and interview them once. Some say that "I am a doctor," a teacher. For example, there was a lady at my house. My father was ill, so she was looking after him. She said, "I am a language teacher there." They don't have anyone who does not have an education, not like ours.

(ML_IW-24)

In most of the cases, this "civilization" emphasis in the male approach can enter into an ugly articulation with sexism. By emphasizing that the women are educated and cultured, the men say that they are superior to their own women.

When you look at it as a man, when you interact with a Georgian lady, you see that there are mountains of differences with the lady at home. Actually they are the same, but she seduces you with her lipstick, makeup, clothing, and such. He could not find what he wanted as a guy from his spouse, mate... When he sees it on her, this seduces him. For example, a woman from Hopa and a woman from Istanbul will get married with an arranged marriage, but they want the same guy. Now when the guy goes to a bake shop, he will first look at the way the girl from Istanbul eats the cake, the way she holds the fork... Then he will look at our girl from Hopa. The girl will immediately take the cake in her hand, or she will salivate I don't know what.

(ML_IW-6)

In a country where women die every day in honor killings and sex work is honorless, almost all of the men interviewed in Hopa speak very highly of the immigrant women who work as sex workers. There are a couple of reasons for that. Perhaps they felt that they had to speak more carefully because a female interviewer spoke to them. But beyond this, none of the male interlocutors said a bad word about these women. Instead they defined them as those "who extend human life" (ML_IW-24). One does need to look for other reasons underlying such perception. These men could ask from the immigrant sex workers things they do not want their women to do. Also, they refer to these women as well educated, saying that these women are university graduates or from prestigious professions in their home countries. In that respect, it is plausible to say that by having women do the things they want, these men perhaps display their manhood. The male interviewees did not even try to hide that they had relationships with these women, often mentioning that their "education," "beauty," and "manners" were indisputable. To get such a woman might have made them to feel a "state of superiority." Of course, their admiration of these women was a positive thing, but the same men spoke of their women with insults. They described their wives who "eat a lot, gain weight, and do not take care of themselves."

Local female perceptions of migrant women

The probability that these women will come across each other in a public space is very low as a result of local women being imprisoned in the house and the women who come from outside being imprisoned in hotels. However, the local women see the effect of the relationships that the men in her family (her spouse, father, or son) establish with these women in every field of their daily lives.

The situation of local women within the family is reshaped by the relationships that are established with these women, as we mentioned above. Even despite this, a large majority of the women from Hopa say that they do this job "because they have to," emphasizing the desperateness of these women.

> I mean, the people who came from there were not welcomed kindly at first because of that prostitution business. A person, without even wanting to, can protest, I mean. Because they were not nice things. Many families have shattered, there were many situations. But I mean there too, who would want to do this job anyway? It is because they really need it that they come here. Sometimes a person thinks about it, that she was forced to it and that the economic situation there is very bad.
>
> (FH_IW-4)

It can be said that the most important tool for the woman who comes from the other side of the border to integrate into life in Hopa has been marriage. Marriage of a migrant woman with a local Muslim man is perceived as a kind of purification among the women even if she did prostitution before. Relationships are established with them, and they act as neighbors.

> If they were my neighbors, I would act as neighbors, but there are none in this environment. Why wouldn't I do it anyway? They are also a human

beings after all. Not all Russians are bad people and doing that kind of business. There are also ones who are good.

(FL_IW-21)

As we see in this section, in Hopa, the point of intersection between the border economy and gender is not the same for local women and the women who come from outside. While the relation of local women with the border economy is established in a mediated form through their spouses' relationship with the economy, the women who come from outside are an important component of the economy as suitcase traders, sex workers, or domestic workers. Yet while the local women are imprisoned in the context that the entertainment sector has created. The women who come from outside are turned into subjects of the same sector, a commodity with which bargains are attempted, and they are imprisoned in hotels.

References

Acar, F., Altunok, G., & Gözdaşoğlu-Küçükalioğlu, E. 2008. "Report Analysing Intersectionality in Gender Equality Policies for Turkey and the EU, QUING Project." Vienna: Institute for Human Sciences. Available online at www.quing.eu/files/results/ir_turkey.pdf.

Archer, R., & Rácz, K. 2012. "Šverc and the Šinobus: Small-Scale Smuggling in Vojvodina." In *Subverting Borders: Doing Research on Smuggling and Small-Scale Trade*, edited by B. Bruns & J. Miggelbrink. Berlin: VS Verlag.

Ashwin, S. 2006. *Adapting to Russia's New Labour Market: Gender and Employment Behaviour*. New York: Routledge.

Bağlı, M., & Özensel, E. 2011. *Türkiye'de Töre ve Namus Cinayetleri: Töre ve Namus Cinayeti İşleyen Kişiler Üzerine Sosyolojik Bir Araştırma*. İstanbul: Destek Yayınları.

Bridger, Sue, Kay, Rebecca, & Pinnick, Kathryn. 1995. *No More Heroines? Russia, Women and the Market*. New York: Routledge.

Delaney, C. 2001. *Tohum ve Toprak*. Istanbul: İletişim Yayınları.

Donnan, H., & Wilson, M. T. 1999. *Borders: Frontiers of Identity, Nation and State*. New York: Berg.

Dudwick, N. 2002. *No Guests at Our Table: Social Fragmentation in Georgia*. In *"When Things Fall Apart: Qualitative Studies of Poverty in the Former Soviet Union*, edited by N. Dudwick, E. Gomart, & A. Marc, with K. Kuehnast. Washington, DC: The World Bank.

Dursun, D. 2007. "Cross-Border Co-operation as a Tool to Enhance Regional Development: The Case of Hopa-Batumi Region" (unpublished thesis). Middle East Technical University.

Erman, T. 1998. Kadınların bakış açısından köyden kente göç ve kentteki yaşam. In *75 Yılda Kadınlar ve Erkekler*, edited by Ayşe Berktay Hacımirzaoğlu. Türkiye İş Bankası Yayınları.

Grillot, C. 2012. "Cross-Border Marriages between Vietnamese Women and Chinese Men: The Integration of Otherness and the Impact of Popular Representations." In *Wind Over Water: Migration in an East Asian Context*, edited by D. W. Haines, K. Yamanaka, & S. Yamashita. New York: Berghahn Books.

Heckman, L. Y., & Aivazishvili, N. 2012. "Scales of Trade, Informal Economy and Citizenship at Georgian- Azerbaijani Borderlands." In *Subverting Borders: Doing Research on Smuggling and Small-Scale Trade*, edited by B. Bruns & J. Miggelbrink. Berlin: VS Verlag.

Kalaycıoğlu, S., & Tılıç, R.H. 1998. "İş ilişkilerine Kadınca Bir Bakış: Ev hizmetinde Çalışan Kadınlar." In *75 Yılda Kadınlar ve Erkekler*, edited by Ayşe Berktay Hacımirzaoğlu. Türkiye İş Bankası Yayınları.

Kandiyoti, D. 1988. "Bargaining with Patriarchy." *Gender and Society* (2) 3: 274–290.

Kandiyoti, D. 2011. *Cariyeler, Bacılar, Yurttaşlar: Kimlikler ve Toplumsal Dönüşümler.* Istanbul: Metis Yayınları.

Kay, R. 2006. *Men in Contemporary Russia: The Fallen Heroes of Post-soviet Change?* Burlington: Ashgate.

Kolivar, A. 2000. *Fadime Kimdir?* Istanbul: Heyamola Yayınları.

Lavigne, Marie. 2007. *The Economics of Transition: From Socialist Economy to Market Economy.* London: Palgrave Macmillan.

Marx, K., & Engels, F. 1998. *Komünist Manifesto ve Komünizmin Ilkeleri.* Ankara: Sol Yayınları.

Mendoza, C. 2009. *Crossing Borders: Women, Migration, and Domestic Works at the Texas-Mexico Divide* (unpublished dissertation). Ann Arbor, MI: University of Michigan.

Najmabadi, A. 2000. "Sevgili ve Ana olarak Erotik Vatan: Sevmek, Sahiplenmek, Korumak." In *Vatan, Millet, Kadınlar*, edited by A. Altınay. İstanbul: İletişim Yayınları.

Özgen, N. 2006. "Öteki'nin Kadını: Beden ve Milliyetçi Politikalar." *Toplum ve Bilim* 19: 125–137.

Shvedova, N. 2009. "Gender Politics in Russia". In *Gender Politics in Post-Communist Eurasia*, edited by Linda Racioppiand Katherine O'Sullivan. East Lansing: Michigan State University Press.

Tverdova, Y. 2011. "Human Trafficking in Russia and Other Post-Soviet States." *Human Rights Review* 12: 3.

Williams, L., & Yu, M. 2006. "Domestic Violence in Cross-Border Marriage—A Case Study from Taiwan." *International Journal of Migration, Health and Social Care* 2 (3/4): 58–69.

Yıldırım, A. 2015. *Sınır-Ötesi Evlilikler ve Sınır Çokkarılılığı: İthal Kuramlar.* Ankara: Ankara Üniversitesi Dergisi.

Conclusion

Intersection of economy, ethnicity, and gender on the Turkey-Georgia border

The postmodern approach to the border, the border regions are defined as "third space," and those living in this spatiality are viewed through identity construction processes with their specific cultures. Thus, concepts such as gender ethnicity, race, and sexual identities become more salient in the social sciences and in the border studies. Also, during this period, the concept of intersectionality introduced by the famous feminist scholar Crenshaw (1994) has come to find a place in border studies, and the inequalities appearing in the intersection of different identities become a significant analytical tool. Following this tendency, in my study, I used this analytical tool to understand the changes of the inter- and intra-group relations of ethnics and gender in the Turkey-Georgia border region.

Shields (2008, 303) explains the theoretical foundation for intersectionality and says that intersectionality "grew from study of the production and reproduction of inequalities, dominance, and oppression." The evolution of intersectionality as a theoretical framework has been traced to Black feminist responses to the limitations of the accumulated disadvantage model (e.g., Mullings 1997; Glenn 1999) and the recognition that the intersection of gender with other dimensions of social identity are the starting point of theory (Crenshaw 1994, 2005). Border as a space of encounter for people of different culture, ethnicity, race, and religion creates a dynamic milieu and hence becomes a very fertile ground for the intersectionalist studies. Gloria Anzaldua (1987, 19), in her book *Borderlands/La Frontera*, uses the concept of the border to refer to the psychological, sexual, and spiritual spaces that are "present wherever two or more cultures edge each other, where people of different races occupy the same territory, where under, lower, middle, and upper classes touch, where the space between two individuals shrinks with intimacy." These "border zones" are understood as areas that "become salient around such lines of sexual orientation, gender, class, race, ethnicity, nationality, age, politics, dress, food, and taste" (Rosaldo 1989, 207–8). Pablo Vila (2005) who works on narrative identity construction on the U.S.-Mexico border, indicates that when discussing the immigrants who live on this border, it is not possible to talk about only one identity. It is necessary to think about different identities of these people like being a woman or man, Catholic or Protestant, worker or entrepreneur besides being an immigrant. For him, "they behave in a particular way because they construct a particular 'narrative identity'" and, in a similar way, Polese (2012, 26) says, "more than their role, it may be interesting to concentrate on the function of borders, and living in borderlands, and do this in a given context, rather than globally. This would mean concentrating on their narratives rather than their real function and how people perceive and report them." Following them, in my study on border regions, I focused on how people perceive and experienced the border in a "given context" of the Turkey-Georgia borderlands.

Border economy is theorized in this study as including types of economic activity unique to itself and as a phenomenon that does not only affect the economic life in the regions where it exists. It also affects the social and cultural life of those regions. The accounts of how the phenomenon of the border economy appeared in Hopa and what kind of developmental trajectory it followed, along with its condition today, is mostly accomplished by the interviews conducted with the locals. The effects of this economy can be most apparently seen in the daily practices of the ethnic groups and gender groups.

The changes that happened in the participation of Hemshins and Lazis in economic life, the changes in their social lives, and the shift in hierarchical relationship that they have with each other allow for a materialist analysis of their relationships. The land ownership is a material historical condition that rendered Lazis the dominant group, especially around the seashore. Their long-standing superiority in the social and political life of the town left its traces in the discourse of segregation, namely the subjective realm. However, as Hemshins started to acquire lands in the seashore or exercised other means of upward social mobility, hegemonic relationships between these two groups started to change. Hemshins and Lazis were equalized in terms of social and political power. The increasing role of Hemshins in the town's economic life changed the capital structure of both groups. However, this situation did not remove the boundaries between Lazis and Hemshinis that come from history; it deepened those that existed and created new discourses of othering. Despite the fact that the border economy created economic relations between different ethnic groups who are living in close proximity to each other, one of the main findings of this study indicates that stereotypes are not eliminated. Indeed, some old prejudices are confirmed, and new ones are formed among these ethnic groups living at the Turkey-Georgia border. The forms in which Hemshins and Lazis have participated in business life with the border economy has caused an unequal and hierarchical structure that continues from the past to the present and deepens in different forms. The economic activities that Hemshins are engaged in are mainly in the entertainment sector: management of hotels, cafes, and discos. Such jobs, on the other hand, are labelled as "second class" and "inferior" in Hopa, mainly by the Lazis due to the cultural understanding. Therefore, even though Hemshins became more powerful economically and started to have a say in the political life of Hopa, they lost prestige because of their engagement in such jobs. Such low prestige is also attributed to the families of Hemshins. Actually, their success in business life creates further stigmatization of Hemshins. This is a new exclusionary and scorning discourse that was mainly produced by Lazis against Hemshins. It appeared with the border economy after the opening up of the border. Hence, new discriminatory processes are deepening the social inequalities between the Lazis and Hemshins.

In Hopa, the division between Hemshins and Lazis seems to have originated from the historical conditions that identify these groups in geographical and occupational terms. Hemshins are identified with agriculture and country life, Lazis with city life and commerce. While Hemshins dealt with agriculture and herding in the villages, Lazis held commerce in their hands at the center of the town. As a result of the rural-to-urban migration, stimulated by the disintegration of agriculture and herding—as well as nationwide modernization practices pursued by the central political authority such as compulsory education—spatial division between these two ethnic groups became less viable. Movement of Hemshins to the center created economic and social tensions that yielded new stereotypes and practices of segregation between these two groups. This process that can be evaluated as an inevitable consequence of the modernization process gained an extraordinary pace after the opening up of border. The most significant consequences of the reopening of the border in 1988 can be summarized as follows. The ethnic composition of the population at the center of Hopa changed, and the economic ascendance of Hemshins also caused material problems in their relations with the Lazis who were the "hosts." Hemshins who became distanced from agriculture advanced rapidly in the field of commerce, migrated to the city center in order to benefit more from education and social opportunities, and the villages became places that are visited in summer for the tea and doing work. It can be said that in terms of ethnic relations in border region, these ethnic groups have different experiences of discrimination and exclusion. After the opening of the Sarp border gate, Lazis lost the economic power that they had for years in Hopa. In the case of Hemshins, however, they gained economic power but lost their prestige, because of their participation in low-prestige jobs (more specifically the business that they carry out in "Teksas Street," which is created by the border economy). Thus, in the case of Hopa, we cannot claim that a single group experienced exclusion and/or discrimination. Economic success does not always bring prestige or eliminate discriminatory processes. Economic disadvantage does not bring low prestige.

Parallel with the above-mentioned new discriminatory ethnic discourse, gender discrimination became deeper after 1988 and the opening of the Sarp border gate. In the case of Hopa, gender relations displayed a multilayered, hierarchical, and intertwined structure. In terms of the relations between local women, there is a process of "othering" that has been constructed on the basis of being a Hemshin or a Lazi woman. Hemshin women who recently moved to the city and remained uneducated could not find a place in the work life, and there were no opportunities except for sitting at the house and "waiting for their husbands," organizing home gatherings (gün), and doing the routine house chores. For Lazi women, an advantage emerged because of living in the city since the beginning, being educated, and as a result, having more chances at work life.

Furthermore, the discriminatory processes against Hemshin men through their business activities also affects their family relations, and Lazi women blame the Hemshin women for accepting the "disgusting" business of their husbands in the entertainment sector. Since they rely on their husbands' earnings, Hemshin women are accused of sharing the economic benefits of the "dirty" sector by Lazi women. For example, a Lazi woman whose husband is a shop owner is more prestigious than a Hemshini woman whose husband works in a nightclub. In addition, from the Lazi perspective, Hemshin women do not leave their husbands even if the latter cheat on them. Another point that Lazi women utter when they justify their perception of their own superiority over Hemshins is based on life in the city. Accordingly, Hemshin women who moved to the city could not act in accordance with the city rules and could not adapt to city life. In daily life, this prejudice manifests itself as not going to each other's houses even though they know each other or not renting their house to Hemshins. Such a hierarchy between Lazi and Hemshin women has a historical background and has deepened with the opening of the border and economic activities. Until the opening of the border, Lazi women were perceived as economically, socially, and culturally superior to Hemshin women. The hierarchy among these women began to change in certain points with the opening up of the border. While the Hemshin women's economic conditions improved with the income their spouses earned from the border trade, they have has started to display themselves in the city life, and they have climbed up in the social structure, this does not break down the social and cultural boundaries and improve their status. Even though they share the same urban space, Lazi women still are "from the city" and Hemshin women are "from the village."

On the other hand, Lazi and Hemshin women together share higher status when compared to immigrant woman. Both local women need to compete with the "other woman" who affected their familial relations. Even though Lazi women argue that the husbands of Hemshin women are involved with immigrant sex workers, the uneasiness that it creates within the family and the probability of disloyal husbands exists for women of both groups. This condition of being in the same situation is also reflected in their attitudes toward the immigrant women. On one hand, Lazi and Hemshin women are angry at the immigrant women, as they are forced to become part of an unfair competition. The women who come from the other side of the border become the common "other" for both groups of local women. Even though they do not meet in the public space with the "well-groomed," "beautiful" other women, a confrontation happens over the relationship their husbands can possibly establish with these women and the discourse that men have about the women For both Hemshin and Lazi men, the arrival of women who are educated and cultured, who know how to talk, and as described by one interviewee as "those who can improve the quality of a person's life" forces their own women to change. According to these men, the Black Sea women who were "eating, drinking, giving birth, and gaining weight" had to "get themselves in shape" in order not to lose their men. On the other hand, the "other" is constantly belittled by the local women of both ethnic groups, emphasizing the "poverty" and "lack of economic privileges" of the immigrant women. Hence they perceive themselves as having much higher status than their immigrant rivals due to their economic position.

In terms of gender relations that change in relation to Sarp border gate, the story of immigrant women is a case of victimization and deprivation, but at the same time, the economic benefits are also significant. The immigrant women were one of the most prominent factors of the economic context that the border created. They were first suitcase traders. When this period ended and there was nothing to sell, they started working as sex workers as a way of escaping poverty in their own countries. Some entered into fake marriages to cross the border legally, and some entered into care work. Almost all of them came from families of the lowest socioeconomic rank in their own countries. However, when working as sex workers, they are seen at the lowest rank in the social hierarchy.

Nevertheless, those women from the other side of the border are a source of sustenance for both enterprise owners who operate hotels and restaurants and for shopkeepers. No matter if they came through a network connection or by themselves, the women had to make a commercial agreement and establish a strategic relationship with the men in Hope to be able to work in more "secure" environments; this secure condition reduces the probability of getting deported. The owners of the enterprises take their share of the earnings of the immigrant women by providing their logistical needs such as lodging and the food and drinks that customers demand. Hence, although economically earning a good income from sex work, they are labelled with the lowest prestige and exploited by hotel owners. The entertainment sector that is brought into being by the border created a context where local women are imprisoned in their houses, migrant women are imprisoned in hotels, and both sides are dependent on men. It can be seen in our case that patriarchal relations create a hierarchy between women and men as well as among women at different levels and forms. The "aggrieved" immigrant women created new losses for the local women. On the other hand, discrimination is also experienced among local women of different ethnic groups. To sum up, a patriarchal, multi-layered, and dynamic hierarchical structure in Hopa emerges both among the local women and the immigrant women as well as among different ethnic groups, based on economic relations. A difficult form of life and patriarchal oppression for women exists in Hopa with the border economy where new types of victimhood and deprivation are multiplying in daily experiences despite the economic opportunities. In terms of gender relations in Hopa, the border economy created much more oppression for all women, but they are oppressed differently, and these differences are based on the participation of women in the border economy and their ethnicity.

The border creates advantages and disadvantages for the people who live around it who are neither merely victims nor glorious beneficiaries. The same border gate can create opportunities from time to time, and it can also cause discrimination and impossibilities. To have the advantages or suffer from the disadvantages depends upon a person's gender or ethnic identity. The gendered and ethnicized structure of daily life allows the intersectionalist analysis of these regions. At the intersection of gender, ethnicity, and economy, identifying winners and losers in local settings has become more difficult and multi-dimensional, and these intersections create an environment where winners and losers are changing all the time.

References

Anzaldua, G. 1986. *Borderlands/La Frontera*. San Francisco: Aunt Lute Books.

Anzaldua, G. 1987. *La Frontera: The New Mestiza*. San Francisco: Aunt Lute Books.

Crenshaw, K. 1991. "Mapping the Margins: Intersectionality, Identity Politics, and Violence against Women of Color." *Stanford Law Review* 43 (6): 1241–1299.

Glenn, E.N. 1999. "The Social Construction and Institutionalization of Gender and Race: An Integrative Framework." In *Revisioning Gender*, edited by Myra Marx Ferree, Judith Lorber and Beth B. Hess. New York: Sage.

Mullings, L. 1997. *On Our Own Terms: Race, Class, and Gender in the Lives of African American Women*. New York: Routledge.

Polese, A. 2012. "Who Has the Right to Forbid and Who to Trade? Making Sense of Illegality on the Polish-Ukrainian Border," in *Subverting Borders: Doing Research on Smuggling and Small-Scale Trade*, edited by B. Bruns and J. Miggelbrink. Berlin: VS Verlag.

Rosaldo, R. 1989. *Culture and Truth: The Re-Making of Social Analysis*. Boston: Beacon.

Shields, A. S. 2008. "Gender: An Intersectionality Perspective." *Sex Roles* 59: 301–311.

Vila, P. 2005. Immigration, Acculturation and Gender Identities on the U.S.-Mexico Border. Paper presented at the annual meeting of the American Sociological Association. Philadelphia.

Index

Pontic culture 15
Pontus Empire 10
poverty 111, 114, 130
price discrimination 39
privatization 21, 22, 23
Procopius 14
prostitution 2, 37, 39, 48, 61, 62–3, 64, 81, 93, 95, 97, 99, 102, 103, 105, 112, 116–18, 119, 120, 121

race 6, 126
Rácz, K. 111
refugees 1
religion 11, 15, 126; *see also* Christians/ Christianity; Islam/Islamization
Rize 9, 12, 16, 18, 19
Roitman, J. 40
Roman Empire 15
Rosaldo, R. 126
Rose Revolution, Georgia (2004) 57
rural-to-urban migration 12, 85, 100, 101, 109–10, 128
Russia 9, 10–11, 16, 38; *see also* Soviet Union
Russia-Turkey border 10–11
Russian bazaars 37, 38, 39–40, 47
Russian Empire 10, 13; and Ottoman conflicts 13, 16
Russians 11, 40, 62, 105, 106, 119, 121, 124

Şahin, Ç. 28, 29, 30, 32
Sakar, M. 50
Sakarya 19
salt 25
Samsun 11, 18
Sana 19
Sapanca 16
Sarıbudak 10
Sarp 10, 11, 14; division into two 24–5
Sarp border gate 5, 12, 23–32; closure of (1937 to 1988) 24, 25, 26–9; opening of (1988) 29–32, 37 (chaos of transition process 31; economic advantages 30, 31, 32; negative perceptions of 30–2)
Şavşat 10, 11, 12
Schendel, W. 23, 24, 49–50, 56, 67
schools, Islamic 10
Scott, J. 64
security 36, 59–60
Selim I 10
Şenkaya district 11
Senoğuz, P. 44, 64, 65

service industry 39
Şevket, Ş. 11
sex industry 60, 62–3, 82; *see also* prostitution
sex tourism 62–3
sex trafficking 94, 97
sex workers 102, 114, 116–18, 120, 121, 122, 124, 130
Seyahatnâmesi, E. C. 9
Shields, A. S. 126
shuttle trade *see* suitcase/shuttle trade
Shvedova, N. 118
Simmel, G. 5
Simonian, H. H. 17, 18, 19, 33n8
small- and medium-scale enterprises 56
smuggling 2, 37, 39, 43, 44, 46, 48, 49, 64–8, 81, 111
social identity 4
social memory 1–2
social welfare 44
Sökmen, Y. O. 14, 33n5
sovereignty, state 3, 4
Soviet Union: border with Turkey *see* Turkey–Soviet Union border; dissolution of 1, 3, 4, 37, 114
"special invoice" document 41–2
Stalin, Joseph 27
structural intersectionality 115
sugar 25
suitcase/shuttle trade 37, 39, 40–2, 43, 47, 112–14, 124, 130
Suleyman Bey 10
Sundura district 74, 75
Sunni Muslims 18
Svan 13, 33n5
Svenatian language 15
Syria 1

Tandilava, A. 14
tax exemption 37, 40, 41
tax regulations 39, 55–6, 66
tea cultivation/collection 12, 20, 21–3, 43, 95–6, 97, 100, 103
"Teksas Avenue" 80, 81–2, 128
third space 3, 126
Tiliç, R.H. 108–9
Tokat 11
Toumarkine, A. 14
tourism 50–1, 62; sex 62–3
Trabzon 9, 10, 11, 16, 18, 19
Trabzon Empire 15
trade agreements, Georgia–Turkey 52
Trade and Economic Cooperation Agreement (Georgia–Turkey, 1992) 52

138   *Index*

Transire (*Pasavan*) document 25
transportation sector 20, 39, 43, 49, 50, 59; Hemshins and 48, 72, 77, 78, 79, 80, 83–4
Trebizond 15
Turkey–Georgia border, reopening of (1988) 4, 5, 24, 29–32
Turkey–Georgia trade agreements 52
Turkey–Russia border 10–11
Turkey–Soviet Union border: closure of (1937 to 1988) 24, 25, 26–9; defined by Treaty of Kars (1921) 24–5; security measures 26–7; tracking field 26–7; transition period (1921 to 1937) 24–6; watchtowers 26
Turkish language 19
Turkish Republic 17, 18, 19, 24, 79, 94
Tverdova, Y. 118

U.S.-Mexican border 4, 119, 126
Üçkardeş 11, 19
unemployment 62, 113, 118
uranium smuggling 44
urbanization 82, 85
Uzbekistan 47

Vanilişi, M. 14
Vaux, B. 18
vegetable oil 25
Viçe (Fındıklı) 10, 12, 15, 16
Voal, P. 126

wealth: accumulation of 43–4; redistribution of 44
Westernization 122

Williams, L. 115
Wilson, T. M. 2, 4, 36, 41, 48, 68, 93
womanization of countries/borders 93–4
women: attitude to border opening 98; competition between local and immigrant 108, 129; Hemshin 97, 108, 109–10, 128–9; hierarchy among 97, 128–9, 130; and household economy 94, 95–6, 97; Lazi 97, 110, 128, 129; legitimization of existence in household 109; loss of social spaces 105–7; and modernization processes 96–7; and the patriarchal bargain 107–8, 109, 110; position within family 96, 101–3; relationship with husbands 99–100; scattered lives of 94–100; tea cultivation/collection 95–6, 97, 103; as victims and beneficiaries 101, 108–10, 130–1; *see also* immigrant women
world of borders 3
World War I 16

Yalova 16
Yoldere/Zhulpiji 19
youth experiences 103–5, 112
Yozgat 11
Yu, M. 115
Yükseker, D. 38, 39, 43, 77
Yusufeli regions 11
Yuval-Davis, N. 6

Zan 33n5
Zeki, M. 12, 19, 20
Zonguldak 12

Milton Keynes UK
Ingram Content Group UK Ltd.
UKHW040051071024
449327UK00019B/478